The Superfamily Mactroidea (Mollusca:Bivalvia) in American Waters

Javier H. Signorelli

The Superfamily Mactroidea (Mollusca:Bivalvia) in American Waters

An Illustrated Catalogue of Recent Species

Springer

Javier H. Signorelli
Instituto de Biología de Organismos Marinos
IBIOMAR, CCT CONICET-CENPAT
Puerto Madryn, Chubut, Argentina

ISBN 978-3-030-29096-2 ISBN 978-3-030-29097-9 (eBook)
https://doi.org/10.1007/978-3-030-29097-9

This Springer imprint is published by the registered company Springer Nature Switzerland AG
The registered company address is: Gewerbestrasse 11, 6330 Cham, Switzerland

Preface

This book is the result of several years of research on the superfamily Mactroidea. I started to study the taxonomy of the species belonging to the family Mactridae of southwestern Atlantic Ocean as the main topic of my PhD dissertation in 2005. One year later, I participated in the International Congress on Bivalvia in the Universitat Autònoma de Barcelona and in the Second International Workshop on Bivalvia in Thailand. It was in that moment that I started to study the described genera and species of the superfamily Mactroidea in a global level. Not only the Recent genera but also the fossil taxa had been studied. All redescriptions are based on the type material and additional specimen examination from more than 15 malacological collections personally examined and from other institutions where the materials were digitally provided. This book constitutes the result of the examination of European and North and South American collections, numerous field works to different localities of South America, and type material examinations of almost 200 nominal species. I am deeply grateful to the curators, technicians and additional people who assisted me during the revision of type materials. This book intends to contribute to the knowledge of the group in American waters and state the bases for future works.

Puerto Madryn, Chubut, Argentina

Javier H. Signorelli

Acknowledgements

I would like to acknowledge all the curators, technicians and additional people who assisted me in the revision of type materials and additional illustrated specimens. They are Gary Rosenberg, Paul Callomon, Jocelyn Sessa and Katy Estes-Smargiassi (ANSP); Elizabeth Kools and Chrissy Garcia (CAS); Konstantin Lutaenko (Far East Branch of the Russian Academy of Sciences); Alejandro Tablado (MACN); Katie Parodi and Adam Baldinger (MCZ); Cristina Damborenea and Gustavo Darrigran (MLP); Fabrizio Scarabino (MNHN); Virginie Heros, Didier Merle and Delphine Brabant (MNHN); Charles Laurent (MHNBx); Sergio Letelier and Catalina Merino (MNHNS); Alexandre Pimenta (MNRJ); Paula Spotorno (MORG); Luis Ricardo Simone and Patricia Oristanio Lima (MZUSP); Jeroen Goud (Naturalis Biodiversity Center); Kathie Way, Andreia Salvador, Amelia MacLellan, Harry Taylor, Phil Hurst and Tom White (NHMUK); Anita Eschner and A. Schumacher (NHMW), Eike Neubert (NMBE); Mark Carnall (Oxford University Museum of Natural History); Paula Mikkelsen and Gregory Dietl (PRI); Anthony Ricciardi and Anthony Howell (Redpath Museum); Boris Sirenko (Russian Academy of Science); Paul Valentich-Scott (SBMNH); Ashley Dineen (UCMP); Ellen Strong, Yolanda Villacampa, Tyjuana Nickens, Ilya Temkin and Mark Florence (USNM); Thomas von Rintelen, Christine Zorn, Matthias Glaubrecht and Frank Koehler (ZMB); and Ole Tendal (ZMUC-BIV). Special thanks to the Malacological Society of London, the Conchologists of America Research Grant, Jessup Travel Award of the Academy of Natural Sciences of Philadelphia and the Ernst Mayr Travel Award Museum of Comparative Zoology, Harvard University. Special thanks to Gregorio Bigatti and to all members of the Laboratorio de Reproducción y Biología Integrativa de Invertebrados Marinos (LARBIM) where I currently develop my research. Final thanks to my advisor Guido Pastorino (MACN) for the many years of training, research and friendship that strongly contribute to my personal and professional development.

Abstract

In this catalogue, the valid species belonging to the superfamily Mactroidea living in American waters are recognized. In the study area, the family Mactridae includes 12 genera and 48 species distributed in three subfamilies (Mactrinae, Lutrariinae and Darininae). There are ten genera belonging to the subfamily Mactrinae living in American waters. They are *Harvella*, *Mactrellona*, *Mactromeris*, *Mulinia*, *Rangia*, *Simomactra*, *Spisula*, *Standella*, *Trinitasia* and *Tumbeziconcha*. Lutrariinae is only represented by the genus *Tresus* and includes three valid species living in the northwestern coast of North America. In addition, the subfamily Darininae is represented by the genus *Darina* with two valid species living in the southern tip of South America. The family Anatinellidae is represented by two genera with two valid species each one, all included into the subfamily Kymatoxinae. Finally, the family Mesodesmatidae is represented in American waters by the genus *Mesodesma* with three valid species and the recently described *Amarilladesma* with only one species.

Keywords Mactroidea, Mactridae, Anatinellidae, Mesodesmatidae, America, Taxonomy

Contents

Chapter 1
Introduction to the Superfamily

Keywords Mactroidea · Mactridae · Anatinellidae · Cardiliidae · Mesodesmatidae · America · Type material

The Superfamily Mactroidea

The superfamily Mactroidea includes 46 described genera and almost 260 species. The supra-generic classification of this superfamily was recently revised and rearranged (Signorelli and Carter, 2016). They recognized eight subfamilies grouped into four families (Table 1.1). Within the family Mactridae four subfamilies are currently included. They are Mactrinae Lamarck, 1809; Lutrariinae Gray, 1853; Darininae Signorelli in Carter et al., 2011 and Tanysiphoninae Scarlato and Starobogatov in Nevesskaja et al., 1971. The latter was described by Scarlato and Starobogatov in Nevesskaja et al. (1971, p. 18) to distinguish the species *Tanysiphon rivalis* Benson, 1858. Although Tanysiphonidae was described as a family level, several authors considered it a subfamily of Mactridae (Boss in Parker, 1982, Bieler et al., 2010; Carter et al., 2011; Signorelli and Carter, 2016). Beu (2006) proposed Zenatinae as a synonym of Lutrariinae. In addition, the subfamily name Kymatoxinae Stenzel and Krause, 1957 [= Pteropsellinae Keen, 1969] was recently relocated to the family Anatinellidae Deshayes, 1853 (Signorelli and Carter, 2016). This family was described to distinguish the genus *Anatinella*, previously introduced by G. B. Sowerby, I (1821–1834), from *Lutraria* Lamarck, 1799.

The family Mesodesmatidae includes two subfamilies: Mesodesmatinae and Davilinae. Mesodesmatidae is a small family with few described species within the genera. Several authors have revised this group (Beu, 1966, 1971; Beu and Rooij-Schuiling, 1982; Davis, 1964, 1965; Lamy, 1912a; L. Reeve, 1854c; Rooij-Schuiling, 1972), though it was Dall (1895a, 1895b, 1898a, 1898b) who strongly contributed to form the base for the subsequent works. The mesodesmatids are

J. H. Signorelli, *The Superfamily Mactroidea (Mollusca:Bivalvia) in American Waters*, https://doi.org/10.1007/978-3-030-29097-9_1

Table 1.1 Revised suprageneric taxonomy of the superfamily Mactroidea

Superfamily	Family	Subfamily	Type genus
Mactroidea	Mactridae	Mactrinae Lamarck, 1809	*Mactra* Linnaeus, 1767
		Lutrariinae J. Gray, 1853	*Lutraria* Lamarck, 1799
		Darininae Signorelli in Carter et al., 2011	*Darina* Gray, 1853
		Tanysiphoninae Scarlato and Starobogatov in Nevesskaja and others, 1971	*Tanysiphon* Benson, 1858
	Anatinellidae	Anatinellinae Deshayes, 1853	*Anatinella* G. B. Sowerby I, 1833
		Kymatoxinae Stenzel and Krause in Stenzel, Krause, & Twining, 1957	*Kymatox* Stenzel and Krause in Stenzel, Krause, & Twining, 1957
	Cardiliidae		*Cardilia* Deshayes, 1835
	Mesodesmatidae	Mesodesmatinae J. Gray, 1840	*Mesodesma* Deshayes, 1832a
		Davilinae Dall, 1895b	*Davila* J. Gray, 1853

characterized by the presence of retractile and naked siphons, not completely fused; cuneiform and compressed shells. Currently, Mesodesmatinae members differ from Davilinae basically in the presence of a pallial sinus and more elongated shells. However, the included species into both subfamilies are similar in porcelaneous texture, with a strong periostracum; hinge plate with a single and narrow cardinal tooth in the left valve, with the posterior arm shorter but extended across the apex of the chondrophore, which is narrow and oblique. Mesodesmatidae was elevated to superfamily status by Yonge and Allen (1985), who based this decision on ligament and siphon morphology. Later, Allen (1985) associated the Mesodesmatoidea with the Tellinoidea. However, Taylor et al. (2007) rejected this relationship after molecular results that highly supported the superfamily Mactroidea as a monophyletic group including Mesodesmatidae.

The superfamily Mactroidea has been studied and documented from several regions, such as Australia and New Zealand (Beu, 1971; Beu and Rooij-Schuiling, 1982; Climo, 1972; Conroy et al., 1993; Darragh and Kendrick, 1971; Gill and Darragh, 1963; Iredale, 1930, 1931; Iredale and McMichael 1962; Kershaw, 1958; Lamprell and Healy, 1998; Marwick, 1952; EA Smith, 1914; Wilson and Kendrick, 1968); Europe (Degraer et al., 2007; Holme, 1959; Jeffreys, 1864; Locard, 1890; Lucas, 1974a, 1974b; Smriglio et al., 1994; Urk, 1959, 1964, 1980); Indo-Pacific (Fischer, 1977; Habe, 1979; Hylleberg et al., 1999; Vongpanich, 2000; Wong, 2009); north-eastern Pacific (Imai et al., 1953; Kessley and Hitz, 1971; Kong and Li, 2009; Lutaenko, 2014; Matsukuma, 2001; Qi, 2004; Zhuang, 1983); Red Sea (Bosch et al., 1995; Lamy, 1916; Oliver, 1992) and western and southern Africa (Bouchet and Nicklès, 1976; Barnard, 1963–1974; Bartsch, 1915; Boshoff and Field, 1968; Branch et al., 2010; Brown, 2002; Von Cosel 1995; Dautzenberg, 1921; Kilburn, 1971, 1973; Kilburn and Rippey, 1982; Nicklès, 1950, EA Smith, 1910; GB Sowerby III, 1892, 1897).

In American waters, several species belonging to this superfamily were widely registered. These groups of bivalves are characterized by middle to large size shell, with a brownish to yellowish periostracum. Internally they present a chondrophore housing the resilium or internal ligament. Finally, the main character that separates this superfamily from the other heterodont bivalves is the typical hinge with the left inverted V-shaped cardinal tooth. Authors like Lamarck (1815–1822), d'Orbigny (1834–1847), Récluz (1852) and Rochebrune and Mabille (1889) described and deposited new species in the Museum National d'Histoire Naturelle of Paris (MNHN). In parallel, in the Natural History Museum of London (NHMUK), authors like Gray (1825, 1837, 1853), Reeve (1854a, 1854b, 1854c), Deshayes (1854, 1855a, b), EA Smith (1885, 1905, 1915) and the clan Sowerby (1825, 1821–1834, 1846), among others, introduced about the 40% of the nominal taxa present in the study area. At the same time, in America, authors like Dall (1891, 1894a, 1894b, 1894c, 1894d, 1895a, 1895b, 1901, 1915), Gould (1841, 1851, 1852–1861), Say (1822) and Philippi (1844–1850, 1845, 1849, 1887a, 1893) described several species for the coast of North America and the southern tip of South America. In the literature, more than 200 names were found living or associated to American forms (Abbott, 1974; Abbott and Dance, 1986; Cantera and Contreras, 1988; Carcelles, 1944, 1950; Carcelles and Williamson, 1951; Castellanos, 1970; Coan and Valentich-Scott, 2012; Coan et al., 2000; Lamy, 1912a, 1912b, 1913, 1914a, 1914b, 1917–1918, 1925; Mikkelsen and Bieler, 2007; Moore, 2002; Rios, 1970, 1975, 1985, 1994, 2009; Huber, 2010, 2015; Saul, 1973, 1991; Scarabino, 2003; Scarabino et al., 2016; Signorelli and Pastorino, 2011, 2012a, 2012b; Signorelli and Carter, 2016; Tryon, 1870; Zelaya, 2016). In this catalogue, all valid species belonging to the superfamily Mactroidea living in American waters are revised and listed.

Morphology of the Group

The class Bivalvia Linnaeus, 1758 is a very diverse group of organisms. They are characterized by the presence of a laterally compressed body, surrounded by two articulated valves. Several names have been used to define this group. However, the term Bivalvia introduced by Buonanni in 1681 and subsequently adopted by Linnaeus (1758) has priority. Other terms include Acephala (Cuvier, 1798), Conchifera (Lamarck, 1818), Pelecypoda (Goldfuss, 1820) and Lamellibranchia (Blainville, 1824).

The members of this class do not have a defined head, cephalic sensory organs or radula. The closure of the valves is produced by the action of the adductors muscles, while the opening is controlled by the ligament, which can be internal and/or external. The main characteristics used to anatomically describe the bivalves are the labial palps, the gills, the stomach and the siphons mainly (Cox in Moore, 1969). Authors such as Atkins (1937), Purchon (1960, 1987), Stasek (1963) and Yonge (1957, 1948), among others, categorized these characters in order to quantify the morphological variation. In addition, the shell characters used to describe the species are mainly the hinge morphology, ligament, escutcheon, sculpture, lunule and umbo.

In general terms, the members of the superfamily Mactroidea are characterized for presenting large to medium size shells, trigonal, oval or elliptical, equilateral with prosogyrate umbos. The hinge plate morphology with the inverted V-shaped cardinal tooth (2a and 2b) in the left valve allows to recognize all mactroids from other superfamilies of the subclass Heterodonta. This tooth is formed by the fusion of two cardinal teeth and, in some species, is flanked by an accessory lamella (4b). Moreover, in the right valve, the cardinal (3a and 3b) teeth are not fused. The anterior and posterior lateral teeth (LAI and LAIII, LPI and LPIII in the right valve; LAII and LPII in the left valve) have usually a single cusp and can be more or less elongated in some taxa. The chondrophore is ventrally projected and well developed in Recent species. The ultrastructure of the shell was studied in different species of the superfamily (Taylor, et al., 1973; Signorelli and Pastorino, 2011, 2012a). They show two aragonitic layers, the external one with a simple crossed structure and the internal one with complex of crossed lamellar structure. The periostracum that covers the shell may be yellowish, reddish or brownish; fine, fibrous and covers the ventral margin of the mantle as well as the siphons. The lunule and the escutcheon are very limited, only observed in some species. Internally, the valves show a conspicuous pallial sinus, in most of the species, which reveal the mode of life of these organisms. In relation to the internal anatomy, the members of Mactroidea show two siphons fused or unfused that can be fully retracted within the shell or not. The adductor muscles are usually the same size. In some genera, (i.e. *Raeta*) the anterior adductor muscle may be more elongated in cross section than the posterior (Harry, 1969). The labial palps are large, well developed, triangular and elongated. The internal surfaces are plicate facilitating the transport of food substances to the mouth. The eulamellibranch ctenidia are formed by two hemibranchs, the external one being usually smaller.

Several publications quantify and classify different species of bivalves according to the anatomy. In relation to the superfamily Mactroidea, Stasek (1963) mentioned that the relationship between the gills and labial palps belongs to the category III. Purchon (1960) classified the stomach as type V. The siphons were also classified according to their morphology. Yonge (1948) includes them in the category C. Additional works have been carried out by using these classifications to revised the functional morphology of included species into the superfamily Mactroidea (Allen, 1975; Barnes and Morton, 1997; Dean, 1980; Huang and Wang, 1995; Ivanov, 1980; Marsh et al., 1976; Morton, 2010; Nakazima, 1965; Narchi, 1981; Paramonova, 1982; Purchon, 1987; Prior, 1974; Reunov et al., 2014; Soyer et al., 1987; Webb, 1987; Yonge, 1948, 1982a).

Biogeographical Provinces

The Recent species included in this catalogue are distributed along 14 biogeographical provinces present in the American marine waters. A detailed biogeographical analysis of the superfamily Mactroidea has not been carried out yet. However, there are some catalogues and checklists that included species belonging to Mactroidea focused on some regions of the American continent (Chamberlin, 1952; Coan, et al., 2000; Coan and Valentich-Scott, 2012; Mikkelsen and Bieler, 2007, Paredes et al., 2016; Rios, 2009; Scarabino et al., 2016; Zelaya, 2016; among others). In this huge area, 16 genera are recognized, all of them, almost exclusive of the region. Only the genera *Simomactra*, *Standella*, *Tresus* and *Raeta* have additional species living in other regions of the world, but the taxonomic placement of those species must be confirmed. Within the American marine waters, there are some genera restricted to a single biogeographical province (i.e. *Mactromeris*, *Amarilladesma*) and other expanded to several provinces (i.e. *Darina*, *Mulinia*, *Standella*, *Spisula*, *Raeta*, *Mesodesma*).

The 14 biogeographical provinces are illustrated in the Fig. 1.1. These areas were mainly introduced by Woodward (1856) and subdivided by subsequent authors (Boschi, 1979, 2000a, 2000b; Scarabino, 1977; Norse, 1993; Roy et al. 1994). Following, the limits of each biogeographical province are listed: (1) Artic. From Aleutian Islands in the Pacific to the Strait of Belle Isle in the North Atlantic Ocean. (2) Aleutian. From Nunivak Island on the northern coasts of the Pacific Ocean to Puget Sound. (3) Oregonian. From Puget Sound to Point Conception. (4) Californian. From Point Conception to Bahia Magdalena. (5) Panamic. From the Isla Cedros—Punta Eugenia area on the Pacific coast of Baja California Peninsula, Mexico to Punta Aguja, Piura, Peru. (6) Peruvian. From Piura, Peru to Chiloe island, Chile. (7) Magellan. From Chiloe Island, Chile to Valdés Peninsula, Argentina. (8) Argentinian. From Valdés Peninsula, Argentina to Cabo Frio, Brazil. (9) Brazilian. From Cabo Frio, Brazil to Orinoco river delta, Venezuela. (10) Caribbean. From Orinoco river delta, Venezuela to Gulf of Mexico, Caribbean Islands, Bermuda Islands and Florida to Cape Canaveral. (11) Texan. From Cabo Rojo to Cape Romano, United States. (12) Carolinian. From Cape Canaveral to Cape Hatteras, United States. (13) Virginian. From Cape Hatteras to Cape Cod, Nova Scotia. (14) Boreal. Extended across the Atlantic from Nova Scotia and Massachusetts to Iceland, the Faeroe and Shetland Islands, the coast of Norway.

Material and Methods

The mentioned materials in this catalogue were meticulously studied. Type material of all nominal species (valid or synonym) mentioned in this catalogue are deposited in the type collections of the following institutions: AMNH—American Museum of Natural History, New York. ANSP—Academy of Natural Science of Philadelphia. CAS—California Academy of Science, Department of Invertebrate Zoology and Geology History. CMNH—Carnegie Museum of Natural History, Pittsburgh. MACN-In—Museo Argentino de Ciencias Naturales "Bernardino Rivadavia", Buenos Aires. MCZ—Museum of Comparative Zoology, Harvard University, Cambridge. MLP—Museo de la Plata; La Plata. MNHN—Museúm National d'Histoire Naturelle, Paris. MHNBx—Muséum d'Histoire naturelle de Bordeaux. MNHNS—Museo Nacional de Historia Natural, Santiago. MNHNM—Museo Nacional de Historia Natural, Montevideo. MNRJ—Museu Nacional de Rio de

Fig. 1.1 Biogeographical provinces of the American continent

Janeiro, Brazil. MORG—Museo Oceanográfico "Prof. E. Rios", Rio Grande do Sul, Brazil. MZUSP—Museu de Zoologia da Universidade de São Paulo, Brazil. NHMD (ZMUC-BIV)—Natural History Museum of Denmark, Zoological Museum, Copenhagen. NHMUK—The Natural History Museum, London. NMBE—Natural History Museum of Bern, Bern, Switzerland. NMNZ—Museum of New Zealand Te Papa Tongarewa, Wellington. PRI—Paleontological Research Institute, Ithaca. RMNH—Rijksmuseum van Natuurlijke Histoire (Naturalis Biodiversity Center), Leiden. SMF Senckenberg Museum, Frankfurt. UCMP—University of California

Museum of Paleontology. USNM—National Museum of Natural History, Smithsonian Institution, Washington DC. ZISP—Zoological Institute, Russian Academy of Science, St. Petersburg. ZMB—Museum für Naturkunde, Berlin. A detailed information of type material repository of all nominal species (valid and synonyms) mentioned in this catalogue is provided in Appendix A.

Descriptions were based on shell and, when it was possible, on soft parts morphology. The hinge was described with the method developed by Bernard and Munier-Chalmas (according to Cox in Moore, 1969) where Arabic numbers are used to designate the cardinal teeth and roman numbers for the lateral teeth. This method shows that all teeth in a right valve have odd numbers and all teeth in a left valve have even numbers. In addition, type locality, type material, geographic distribution, stratigraphic distribution, synonyms, remarks and relevant literature are provided for each valid species. All supra-specific taxa were properly diagnosed. The synonyms mentioned in this catalogue were checked and confirmed after the examination of the type material and original illustrations of all related nominal species.

Chapter 2
The Subfamily Mactrinae (Mollusca: Bivalvia), Part 1

Abstract Mactrinae is the largest subfamily of this group of bivalves. In the American continent 43 species grouped into ten genera are recognized. In this chapter, the species included in the genera *Harvella*, *Mactrellona*, *Mactromeris*, *Mulinia* and *Rangia* are redescribed. The genus *Harvella* includes the only one living species *H. elegans* (G.B. Sowerby I, 1825). The genus *Mactrellona* Marks, 1951 groups four species *M. alata* (Spengler, 1802), *M. clisia* (Dall, 1915), *M. exoleta* (Gray, 1837) and *M. subalata* (Mörch, 1861). The genus *Mactromeris* Conrad, 1868 includes the Pacific species *M. catilliformis* (Conrad, 1867), *M. hemphillii* (Dall, 1894a) and *M. polynyma* (Stimpson, 1860). The genus *Mulinia* is one of the most represented ones with nine living species in the American continent. They are *M. bicolor* Gray, 1837, *M. byronensis* Gray, 1837, *M. cleryana* (d'Orbigny, 1846), *M. edulis* (King, 1832), *M. exalbida* Gray, 1837, *M. lateralis* (Say, 1822), *M. levicardo* (E. A. Smith, 1881), *M. modesta* (Carpenter, 1864) and *M. pallida* (Broderip and G.B. Sowerby I, 1829). The genus *Rangia* Des Moulins, 1832 with four living species is the last group revised in this chapter. *Rangia* (*Rangia*) *cuneata* (Sowerby I, 1832, in 1821–1834), *Rangia* (*Rangianella*) *flexuosa* (Conrad, 1840), *Rangia* (*Rangianella*) *mendica* (Gould, 1851) and *Rangia* (*Rangianella*) *pontchartrainensis* (Morrison, 1965) (new combination) belong to this genus.

Keywords *Harvella* · *Mactrellona* · *Mactromeris* · *Mulinia* · *Rangia*

CLASS BIVALVIA LINNAEUS, 1758
ORDER CARDIIDA FÉRUSSAC, 1822 IN FÉRUSSAC ET AL, 1821–1822
SUPERFAMILY MACTROIDEA LAMARCK, 1809

Family Mactridae Lamarck, 1809

Shell aragonitic, ultrastructure with two layers, the outer crossed-lamellar layer and the inner complex crossed-lamellar layer, equilateral to subequilateral, umbos prosogyrate, external surface without sculpture or with commarginal ribs; external ligament small, internal ligament placed on chondrophore ventrally projected; heterodont hinge plate;

J. H. Signorelli, *The Superfamily Mactroidea (Mollusca:Bivalvia) in American Waters*, https://doi.org/10.1007/978-3-030-29097-9_2

left valve with bifid cardinal tooth; right valve with cardinal teeth not fused or partially united; lateral teeth generally present in both valves; accessory lamellae usually present; mantle lobes not extensively fused, siphons fused, ctenidia synaptorhabdic, homorhabdic, rarely heterorhabdic. Stomach type 5 of Purchon, style sac in a separate caecum.

Subfamily Mactrinae Lamarck, 1809

[nom. correct. H. Adams and A. Adams, 1856, pro subfamily Mactrina, nom. transl. et correct. Gray, 1853, p. 41 ex family les Mactracées [vernacular] Lamarck, 1809, p. 318, latinized by Children, 1822, p. 86, as Mactracea] [= Rangiidae P. Fischer, 1884, p. 118, proposed as a replacement name for Gnathodontidae, nom. correct. P. Fischer, 1884, p. 118 pro Gnathodontiden [vernacular] Neumayr, 1884, p. 408, invalid, based on *Gnathodon* Gray in G. B. Sowerby I, 1832, in 1821–1834] [= *Rangia* Des Moulins, 1832], a junior homonym of *Gnathodon* Goldfuss, 1820, and *Gnathodon* Oken, 1816 (in Oken, 1815–1816) [Pisces], the latter being not an available name, being published in a work placed on the Official Index. Art. 40.2 does not change the year of precedence of Rangiidae because Gnathodontidae was published in the same year.

Shell subequilateral, trigonal, not widely gaping; right valve with two cardinal teeth, not fused, divergent; left valve with strong, inverted V-shaped cardinal tooth, lateral teeth well developed.

Genus Harvella *Gray, 1853*

Shell subequilateral, sub-trigonal; with heavy commarginal undulations, internally observed; external and internal ligament separated by shelly ridge.

Type species: *Mactra elegans* G. B. Sowerby I, 1825, by monotypy.

Geographic distribution: west coast of America. The genus is known to the Miocene.

Remarks: shell outline similar to *Mactrellona*, distinguished by the commarginal sculpture in the external surface. *Harvella elegans* is the only one recognized living species. It was registered from Neogene deposits of west North, Central and west South America (Anderson, 1929; Olsson, 1964; Woodring, 1982). *Harvella sincola* (Olsson, 1922) was recorded from Miocene deposits of Mexico and Pliocene deposits of Costa Rica (Woodring, 1982; Perrilliat, 1992).

Harvella elegans (G. B. Sowerby I, 1825) (Fig. 2.1a–d)

Shell ovate, slightly elongated, inflated, fragile, thin, translucent; umbos prosogyrate; external surface strongly sculptured by commarginal undulations, internally evidenced; antero-dorsal margin concave and dorsally prolonged, postero-dorsal

Fig. 2.1 (**a–d**) *Harvella elegans* (G. B. Sowerby I, 1825), NHMUK 20190673, (**a**, **c**) /1, (**c**, **d**) /2, syntypes; (**e–h**) *Mactrellona alata* (Spengler, 1802), ZMUC-BIV, holotype. Scale bar (**a–d**) 1 cm, (**e–h**) 3 cm

area narrow; anterior end rounded; posterior defined by a keel and somewhat gaped; periostracum brownish and thin; hinge with lateral teeth short closely placed to the cardinals, trigonal chondrophore ventrally projected; pallial sinus not evident, moderate in depth, camouflaged by the external ornamentation.

Type locality: unknown, not mentioned in the original description.
Type material: NHMUK 20190673/1, /2, two syntypes.
Geographic distribution: western America from Gulf of California to Peru.
Stratigraphic distribution: Miocene—Recent.

Synonyms: *Harvella pacifica* Conrad, 1867: 192; *Mactra estrellana* Olsson, 1922: 435; *M. (Harvella) sanctiblasii* Maury, 1925: 114; *Raeta maxima* Li, 1930: 263; *Harvella elegans tucilla* Olsson, 1932: 129.

Remarks: Of all species mentioned as synonyms in the literature, only *Harvella pacifica* (ANSP 51308, syntype) and *Raeta maxima* (AMNH 268093, lectotype; AMNH 268093a, paralectotype) are based on Recent specimens living in Panama bay. *Mactra estrellana* (PRI 21265, holotype) was described from Middle Miocene deposits of Estrella River, Costa Rica. *Mactra (Harvella) sanctiblasii* (PRI 2354, holotype) was described from Miocene deposits exposed near the southern coast of Trinidad. Finally, *Harvella elegans tucilla* (PRI 2252, holotype) was registered in Upper Miocene deposits exposed at Quebrada Tucillal, Tumbes department, Peru.

Relevant literature

Fisheries: Ruiz et al., 2012: 22.

Paleontology: Hendy et al., 2015: 53; Olsson, 1961: 329–330; Pilsbry and Olsson, 1941: 73; Woodring, 1982: 652–653.

Taxonomy: Bernard, 1983, 39; Cauquoin, 1969:1027–1028; Coan and Valentich-Scott, 2012: 545; Dall, 1894c: 28; Hertlein and Strong, 1950: 235; Huber, 2010: 446; Keen, 1971: 206; Paredes et al., 2016: 140; Ramírez et al., 2003, 268; Signorelli, 2012: 195.

Genus Mactrellona *Marks, 1951*

Shell trigonal, thick, large, inflated, umbos prominent, external surface smooth, with commarginal striae, postero-dorsal area defined by a keel; posterior lateral teeth short.

Type species: *Mactra alata* Spengler, 1802, by original designation.

Geographic distribution: Atlantic and Pacific coast of America.

Remarks: This genus was introduced by Marks (1951) to group three species: *M. alata* (Spengler, 1802), *M. exoleta* (Gray, 1837) and *M. clisia* (Dall, 1915). Historically, these species were included in *Mactrella* Gray, due to several authors erroneously considered *M. alata* Spengler as its type species (Chenu, 1859–1862; Conrad, 1868; Dall, 1915; Lamy, 1917–1918; among others). However, Gray (1853) assigned *Mactra striatula* Linnaeus, 1767, as type species of *Mactrella*. The diagnosis of *Mactrella* given by Gray fits more adequately to *M. alata* than to *M. striatula*. This problem was treated by Marks (1951) who accepted the designation of *M. striatula* as type of the genus *Mactrella* and described a new genus. In this catalogue, four species are recognized. However, a more detailed comparison between *M. alata*, *M subalata*, and related synonyms is needed.

Mactrellona alata (Spengler, 1802) (Fig. 2.1e–h)

Shell trigonal, inflated, inequilateral; umbos prominent, prosogyrous; external surface smooth; anterior end rounded, ventral margin convex; posterior area typically defined by a keel-like ridge; greyish, thin periostracum; internally white, left hinge with the diagnostic V-shaped cardinal tooth and short anterior and posterior lateral teeth; right hinge with two anterior and two posterior lateral teeth, dorsal one smaller, cardinal teeth unfused in this valve; chondrophore well developed, flanked dorsally by a shell ridge which separates the external ligament from the resilium; valves slightly gaping at the posterior end; ultrastructure consisting of an outer crossed-lamellar and an inner complex crossed-lamellar structure; pallial sinus deep, about 50% shell length.

Type locality: "Sudamérica" only mentioned in the original description.

Type material: ZMUC-BIV holotype, articulated specimen; ZMUC-BIV, paratype, one complete specimen.

Geographic distribution: From Gulf of Mexico to Santa Catarina, Brazil.

Stratigraphic distribution: Miocene—Recent.

Synonyms: *Mactra laevigata* Schumacher, 1817: 167; *Mactra carinata* Lamarck, 1818: 473; *Mactra concentrica* Bory de Saint Vincent, 1827: 151 (in Bruguière et al., 1789–1832).

Remarks: *Mactra carinata* (MNHN-IM-2000-7038, syntypes) is confirmed as a junior synonym of the western Atlantic *Mactrellona alata*. The examination of type material and original illustration of *Mactra concentrica* Bory de Saint Vincent, 1827 and *Mactra laevigata* Schumacher, 1817 do not confirm them as synonyms of *M. alata* or *M. subalata*. *M. concentrica* and *M. laevigata* are based on the same figure of the "Tableau of Encyclopedique Methodique ..." (pl. 251, fig. 2a-b, pl. 252, fig. 2c) published in 1797 without a nomenclatorial act. No locality was provided. The shell shape and posterior end of these figures fit more adequately to *M. subalata* than to *M. alata*. The synonymy with *M. alata* is maintained, awaiting additional analysis.

Relevant literature

Ecological and Environmental studies: De Castro and Dos Santos, 1989: 103; Gondim et al., 2011: 5; Passos and Magalhães, 2011: 148.

Fisheries: Carpenter, 2002: 60; Pilsbry and Olsson, 1941: 73.

Paleontology: Maury, 1917: 231; Olsson, 1922: 435; 1961: 327; Woodring, 1982: 649.

Taxonomy: Adams and Adams, 1856 in 1853–1858: 377; Cauquoin, 1969: 1026; Conrad, 1868: 34; Dall, 1894c: 26, 1898a: 877; Gundlach, 1883b: 482; Huber, 2010: 446; Lamy, 1917: 264; Reeve, 1854a: sp 29; Rios, 1975: 236, 1994: 266, 2009: 537; Robinson and Montoya, 1987: 396; Signorelli and Pastorino, 2012b: 36.

Fig. 2.2 (**a**, **b**) *Mactrellona clisia* (Dall, 1915), USNM 271481, holotype; (**c**–**f**) *Mactrellona exoleta* (Gray, 1837), NHMUK 196327, lectotype; (**g**, **h**) *Mactrellona subalata* (Mörch, 1861). Scale bar (**a**–**f**) 3 cm, (**g**, **h**) 2 cm

Mactrellona clisia (Dall, 1915) (Fig. 2.2a, b)

Shell trigonal, medium to large, length to 100 mm, thin; external surface smooth with concentric irregular growth lines; anterior end rounded, antero-dorsal margin narrowed and depressed; postero-dorsal margin straight to slightly concave defined by a winged keel, posterior gape moderate; postero-ventral margin bulged; periostracum thin, yellowish, translucent, absent in the umbonal area; internally white with a hinge plate anteriorly inclined, lateral teeth elongated and cardinal teeth placed on the anterior side of resilium; chondrophore not ventrally projected; pallial sinus moderately deep, about 30% of shell length.

Type locality: Manzanillo, West Mexico.
Type material: USNM 271481, Holotype, articulated specimen.
Geographic distribution: western tropical America from Sonora, Mexico to Máncora, Peru.
Stratigraphic distribution: Recent.
Synonyms: None.
Remarks: *Mactrellona clisia* is a valid species described by Dall (1915). However, anatomy, reproduction or fisheries studies on this species were not carried out.
Relevant literature

Taxonomy: Cauquoin 1969: 1025–1026; Coan and Valentich-Scott, 2012: 545–547; Dall, 1915: 62–63, 1916b: 415; Hertlein and Strong, 1950: 233–234; 1955: 204–205; Huber, 2010: 446; Keen, 1971: 206; Olsson, 1961: 328; Paredes et al., 2016: 141; Signorelli, 2012: 196.

Mactrellona exoleta (Gray, 1837) (Fig. 2.2c–f)

Shell trigonal; large, equivalve, subequilateral, inflated, length to 135 mm; postero-ventral margin evenly rounded; low radial ridge separating central and posterior slopes, without winged keel; postero-dorsal area flat and wide; posterior gape long, narrow; external surface marked by coarse growth lines and covered by a thin, yellowish to brownish periostracum; internally white; hinge plate anteriorly inclined, with lateral teeth closely placed to the cardinal teeth, anterior lateral teeth more elongated than posterior; chondrophore strong subcircular to trigonal; pallial sinus V-shaped, deep; adductor muscle scar subcircular to slightly elliptical, similar in size.

Type locality: Not stated.
Type material: NHMUK 196327, syntype, one complete specimen.
Geographic distribution: Pacific coast of America from Gulf of California, Mexico to Máncora, Peru.
Stratigraphic distribution: Miocene of Peru and Costa Rica—Recent.

SYNONYMS: *Lutraria ventricosa* Gould, 1851: 89 (invalid: junior homonym of *Lutraria ventricosa* Goldfuss, 1840 [in Goldfuss (1833, 1835, 1836, 1837, 1841]).

REMARKS: *Lutraria ventricosa* Gould, 1851 (MCZ 169451, holotype; MCZ 169452, paratype; USNM 15902, paratypes, 4 valves) is confirmed as a junior synonym of *M. exoleta*. The holotype was correctly illustrated by Johnson (1964, pl. 27, fig. 3).

RELEVANT LITERATURE

Ecological and Environmental studies: Brusca and Thomson, 1975: 44.

Paleontology: Grant and Gale, 1931: 402; Marks, 1951: 357; Olsson, 1922: 434; Packard, 1916: 281–282.

Taxonomy: Cauquoin, 1969: 1025–1026; Coan and Valentich-Scott, 2012: 547; Hertlein and Strong, 1950: 234; Huber, 2010: 446; Keen, 1971: 206–207; López de Mesa and Cantera, 2015: 8; Olsson, 1961: 328; Paredes et al., 2016: 141; Signorelli, 2012: 196; Strong, 1925.

Mactrellona subalata (Mörch, 1861) (Fig. 2.2g, h)

Shell trigonal, equivalve, subequilateral, large, shell length to 85 mm; external surface smooth; ventral margin curved; posterior end with a weak keel; postero-dorsal area flat; posterior gape absent; periostracum light brown, eroded in the umbonal area; internal surface white; hinge plate similar to other *Mactrellona* species; chondrophore slightly projecting ventrally; pallial sinus relatively shallow.

TYPE LOCALITY: Not stated.

TYPE MATERIAL: ZMUC-BIV, one syntype.

GEOGRAPHIC DISTRIBUTION: Pacific coast of America from Nayarit, Mexico to Tumbes, Perú.

STRATIGRAPHIC DISTRIBUTION: Recent.

SYNONYMS: None

REMARKS: Signorelli and Pastorino (2012b) considered *Mactra subalata* as a junior synonym of *Mactrellona alata*. However, Huber (2010) and Coan and Valentich-Scott (2012) treated Mörch species as the fourth valid species distributed along the eastern Pacific. After the re-examination of types, original illustrations and additional records, *Mactrellona subalata* is herein accepted.

RELEVANT LITERATURE

Ecological and Environmental studies: Esqueda-González et al., 2014: 52.

Taxonomy: Cauquoin, 1969: 1026–1027; Coan and Valentich-Scott, 2012: 547; Hertlein and Strong, 1950: 234; Huber, 2010: 446; Keen, 1966: 10, 1971: 206–207; Krebs, 1864: 105; López de Mesa and Cantera, 2015: 8; Olsson, 1935: 105, 1961: 328; Paredes et al., 2016: 141; Poutiers, 1995: 145; Signorelli and Pastorino, 2012b: 36–37.

Genus Mactromeris *Conrad, 1868*

Shell large, thick, subequilateral anteriorly elongated, postero-dorsal area defined by a shelly ridge; eternal surface smooth; external and internal ligaments not separated; chondrophore ventrally projected; short lateral teeth, pallial sinus U-shaped, broad.

TYPE SPECIES: *Mactra ovalis* Gould, 1841 (*non* J Sowerby, 1817) (= *Mactra polynyma* Stimpson, 1860, replacement name), by subsequent designation Stoliczka, 1871(in 1870–1871).

GEOGRAPHIC DISTRIBUTION: North Pacific and North Atlantic Ocean.

REMARKS: This genus currently groups three living species: *M. catilliformis* (Conrad, 1867), *M. hemphillii* (Dall, 1894a) and *M. polynyma* (Stimpson, 1860). *Mactra polynyma* is a replacement name for *Mactra ovalis* Gould, 1841 (*non* J Sowerby, 1817, Great Britain). This is the type species of the genus. Johnson (1964) mentioned that the type material of Gould is probably lost. The analysis of related types and additional material suggests that this group needs more work. The three species differ considerably in shell outline. However, Packard (1916) suggested that Recent species are more variable in the convexity of antero-dorsal area. Moreover, the taxonomic status of the circumpolar *M. polynyma* must be revised. As suggested by Coan et al. (2000) and Huber (2010), it might be possible that more than only one species are represent along its wide distribution range.

Mactromeris catilliformis (Conrad, 1867) (Fig. 2.3a–d)

Shell large, ovate to trigonal, inflated, length to 150 mm, inequilateral; umbos small; lunule and escutcheon narrow; slightly longer posteriorly, postero-dorsal margin almost straight, set off by a shallow sulcus, posterior end rounded, gaping, anterior end rounded; external surface with prominent growth lines covered by a gray and wrinkled periostracum; internally with, hinge plate concentrated, lateral teeth short and chondrophore oblique, large and ventrally projected; pallial sinus deep, U-shaped, about 50% of shell length.

TYPE LOCALITY: Panama, but according to Dall (1894a), this locality is incorrect.

TYPE MATERIAL: Not found at ANSP cabinets.

GEOGRAPHIC DISTRIBUTION: West coast of North America from Northern California, United States to Baja California, Mexico.

STRATIGRAPHIC DISTRIBUTION: Miocene of North America—Recent.

SYNONYMS: *Standella californica* of authors (*non* Conrad, 1837); *Spisula catilliformis alcatrazensis* Arnold, 1907: 437; *Spisula mercedensis* Packard, 1916: 286.

REMARKS: This is a well-known species distributed along the west coast of North America. Coan et al. (2000) suggested that *Hemimactra lenticularis* Gabb, 1866 (ANSP 4549, lectotype designated by Stewart, 1930: 208) is an older name for

Fig. 2.3 (**a–d**) *Mactromeris catilliformis* (Conrad, 1867), NHMUK catalogue numbers not assigned, specimen (*non* type); (**e–h**) *Mactromeris hemphillii* (Dall, 1894a), USNM 15815, holotype; (**i–l**) *Mactromeris polynyma* (Stimpson, 1860), AMNH 34018. Scale bar (**a–l**) 3 cm

this species. In addition, two fossil species have been mentioned in the literature as junior synonyms of *M. catilliformis*. They are *Spisula catilliformis alcatrazensis* (USNM MO 165291, holotype) from Pliocene of Fernando Formation deposits exposed at Santa Barbara, California. And *Spisula mercedensis* (UCMP 12067, holotype) from Pliocene deposits of San Mateo County, California.

Relevant literature

Ecological and Environmental studies: Hernández-Alvarez et al. 2013: 201.
Paleontology: Arnold, 1907: 437; Grant and Gale, 1931: 398–399; Hertlein and Grant, 1972: 313–314; Packard, 1916: 285–286; Reagan, 1909: 212; Stewart, 1930: 208.
Taxonomy: Coan and Valentich-Scott in Carlton, 2007: 848; Coan and Valentich-Scott, 1997: 17; Coan et al., 2000: 454; Dall, 1894a: 137, 1916a: 40; Strong, 1925.

Mactromeris hemphillii (Dall, 1894a) (Fig. 2.3e–h)

Shell trigonal to sub-ovate, large, length to 165 mm; inflated, subequilateral to longer anteriorly; umbos prominent; antero-dorsal margin concave, anterior end rounded; postero-dorsal slope convex, set off by low, shelly ridge, with irregular wrinkles; posterior end not gaping; external surface smooth with irregular and commarginal undulations covered by a brownish periostracum; internally white; hinge plate with lateral teeth short and finely striae, chondrophore ventrally projected; pallial sinus U-shaped moderately deep, about 40% shell length.

Type locality: San Diego, California, United States.
Type material: USNM 15815, syntype, two single valves; CAS 66541, probably syntype, specimen from Hemphill collection, collected at type locality.
Geographic distribution: From Central California, United States to southern Baja California, Mexico.
Stratigraphic distribution: early Miocene of North America—Recent.
Synonyms: *Spisula camaronis* Dall, 1921b: 22; *Spisula hemphillii orcutti* Manger, 1934: 301.
Remarks: Two fossil species are commonly cited in the literature as synonyms of *M. hemphillii*. They are *Spisula camaronis* (USMN MO 333117, syntype) from Pliocene deposits exposed at San Quentin bay. And *Spisula hemphillii orcutti* (types not found) from the same locality. Hertlein and Grant (1972) doubted about the synonymy due to outline shell variation. In addition, Hertlein and Grant (1972) mentioned two other fossil species very similar to *M. hemphillii*. They are *Spisula mossbeachensis* Glen, 1959 (UCMP 37643, holotype, UCMP 37644–37646, paratypes), a Pliocene species from the lower portion of the Merced Formation, and *S. longa* Dall, 1921b (USNM 333116, syntype) from the Pleistocene at San Quintin. However, the latter is a junior synonym of *Tresus nuttallii* (Conrad, 1837).
Relevant literature

Paleontology: Dall, 1921b: 22; Glen, 1959: 175; Grant and Gale, 1931: 398; Hertlein and Grant, 1972: 315–316; Packard, 1916: 287; Powell et al., 2005: 186; Valentine, 1957: 297.

Taxonomy: Coan and Valentich-Scott in Carlton, 2007: 848; Coan and Valentich-Scott, 1997: 17; Coan et al., 2000: 454; Dall, 1894a: 137, 1916a: 40; Strong, 1925.

Mactromeris polynyma **(Stimpson, 1860) (Fig. 2.3i–l)**

Shell large, thick, length to 115 mm; moderately compressed, sub-oval, subequilateral; umbos small, not inflated; antero-dorsal margin slightly concave, anterior end rounded; postero-dorsal slope convex, posterior end narrowly gaped; external surface with coarse growth lines covered by a thick, strong, dusky-brown periostracum; internally bluish white; hinge plate with strong cardinal tooth, lateral teeth short, not striated; chondrophore slightly ventrally projected; pallial sinus moderate.

TYPE LOCALITY: Bank fisheries, Boston Harbor, United States mentioned by Gould (1841) in the original description of *Mactra ovalis*.

TYPE MATERIAL: specimens in the stomachs of fish in Boston Harbor. Mentioned as being in Massachusetts State Cabinet No. 229 and Boston Society of Natural history (BSNH) No. 2324 but both specimens are lost (Johnson, 1964).

GEOGRAPHIC DISTRIBUTION: Circumboreal, Bering Sea to Kamchatka, Puget Sound, Greenland to Rhode Island, United States.

STRATIGRAPHIC DISTRIBUTION: Pliocene of western North America—Recent.

SYNONYMS: *Mactra similis* Wood, 1828: 4 (*non* Say, 1822: 309); *Mactra grandis* Deshayes, 1832: 395 (*non* Gmelin, 1791: 3259); *Mactra ovalis* Gould, 1841: 53 (*non* J Sowerby, 1817: 136); *Mactra ponderosa* Philippi, 1844, in 1844–1850:165 (no Eichwald, 1830: 207); *Mactra grayana* Schrenk, 1867: 572; *Callista voyi* Gabb, 1869: 24; *Mactra loebbeckeana* Weinkauff, 1881: 60 (in Küster et al., 1841–1884); *Spisula alaskana* Dall, 1894d: 40 (*nomen nudum*), validated in Dall (1921a: 51); *Spisula vladivostokensi* Bartsch, 1929: 139 and *Spisula korolevae* Zhidkova et al., 1972: 140.

REMARKS: *Mactra polynyma* is a replacement name for *Mactra ovalis* Gould, 1841 (*non* J Sowerby, 1817). The presence of several nominal species mentioned as synonyms of *M. polynyma* suggests that this species complex must be revised. A first approach was done by Cassista and Hart (2007) who analyzed genetically different populations of *M. polynyma*. The study showed genetic similarities among northwest Atlantic populations, and significant spatial differentiation only on the largest geographic scale (between Atlantic and Pacific populations). The type locality of *Mactra similis* and *M. loebbeckeana* was not stated. *Mactra ovalis*, *M. grandis* (MNHN catalogue number not assigned, probable holotype) and *M. ponderosa* were described from North Atlantic localities, whereas *Mactra grayana*, *Spisula alaskana* and *S. vladivostokensis* were described from North Pacific, the latter from The Peter the Great Gulf, Russia. Finally, two fossil nominal species have been treated as synonyms of *Mactromeris polynyma*. They are *Callista voyi* (UCMP 12993, holotype) Pliocene registered near Humboldt Bay, United States and *Spisula korolevae* (types not found) from Neogene of Kurile islands.

Relevant literature

Anatomy and Morphology: Kellogg, 1915: 647–648.
Ecological and Environmental studies: Cassista and Hart, 2007; Chamberlin and Stearns, 1963; Smith and Wikfors, 1992; Steele et al., 1979.
Fisheries: Amaratunga and Rowell, 1988; Davis and Shumway, 1996; Feder et al., 1976, 1977, 1979.
Paleontology: Grant and Gale, 1931: 394–395; Packard, 1916: 283–285; Stewart, 1930: 207.
Taxonomy: Abbott and Dance, 1986: 335; Bernard, 1983: 40; Brunel et al., 1998: 128; Coan and Valentich-Scott, 1997: 17; Coan et al., 2000: 456; Dall, 1894c: 26, 1916a: 40;
Gould, 1870: 75–77; Huber, 2010: 444; Okutani et al., 1989: 120; Scarlato, 1981: 399–400; Stimpson, 1860: 3; Strong, 1925; Tsuchida and Kurozumi, 1995: 23.

Genus **Mulinia** *Gray, 1837*

Shell trigonal to subcircular, medium size to large, length to 12 cm, external surface smooth with commarginal striae; ligament exclusively internal; pallial sinus short, not gaped.

Type species: *Mulinia typica* Gray, 1837 (= *Mactra edulis* King, 1832), by original designation.

Geographic distribution: Pacific and Atlantic coasts of North, Central and South America.

Remarks: The main character that defines this genus is the absence of external ligament. It is well represented in the southern tip of South America with five species. They are the type species *Mulinia edulis* (King, 1832), *M. bicolor* Gray, 1837, *M. exalbida* Gray, 1837, *M. byronensis* Gray, 1837 and *M. levicardo* (E. A. Smith, 1881). However, additional species included in this genus have been registered in other regions. Along the west coast of United States and Mexico, *M. modesta* (Carpenter, 1864) is present. From Mexico to northern Peru *M. pallida* is commonly registered. On the other hand, along eastern coast of North America, *M. lateralis* has been mentioned in the literature (Mikkelsen and Bieler, 2007). Signorelli and Pastorino (2011) expanded the geographic distribution of the genus *Mulinia* to East Africa based on the species *Mactra rodatzi* Dunker, 1850. This species was placed in *Mulinia* by Lamy (1917–1918). Recently, Huber (2010) mentioned that *M. rodatzi* was precisely illustrated in Dunker's Novitates Conchologicae and suggested that the type locality is wrong (See remarks of *Mulinia edulis*). The type material of Dunker was not found.

Mulinia bicolor Gray, 1837 (Fig. 2.4a, b)

Shell trigonal, subcircular; maximum length to 122.5 mm; anterior and posterior ends rounded; ventral margin curved. External surface smooth, with yellow to brownish periostracum; internally white; right hinge with two anterior (AI and AIII) and two posterior (PI and PIII) lateral teeth; anterior teeth similar in length, dorsal tooth stronger than ventral one; ventral posterior lateral tooth more elongated than

Fig. 2.4 (**a**, **b**) *Mulinia bicolor* Gray, 1837, NHMUK 196323, holotype; (**c**–**h**) *Mulinia byronensis* Gray, 1837 NHMUK 196324, syntypes. Scale bar (**a**, **b**) 4 cm, (**c**–**f**) 2 cm, (**g**, **h**) 1 cm

dorsal; two unfused cardinal teeth (3a and 3b) flanked by accessory lamella; left valve with one elongated anterior lateral tooth (AII); posterior lateral tooth (PII) shorter; cardinals fused, V-shaped.

TYPE LOCALITY: Unknown, not indicated in the original description.
TYPE MATERIAL: NHMUK 196323, holotype, one left valve.
GEOGRAPHIC DISTRIBUTION: Caldera to Talcahuano, Chile.
STRATIGRAPHIC DISTRIBUTION: Recent.
SYNONYMS: None.
REMARKS: *Mulinia bicolor* is one of the largest species of the genus (holotype length: 12.25 cm). It is distributed along south-central coast of Chile (Soot-Ryen, 1959; Osorio et al. 1979; González et al. 2002). Despite its large size, it is a very rare species judged by its presence in the collections studied. Gray (1837) included this species in a group with the anterior end delimited by a line. Signorelli and Pastorino (2011) revised morphologically this species by the examination of type material. No synonyms were recognized.
RELEVANT LITERATURE

Fisheries: Osorio et al., 1979.
Taxonomy: Bernard, 1983: 40; Coan and Kabat, 2017: 92; d'Orbigny, 1846 (in 1834–1847): 510; Dall, 1909: 274; González et al., 2002; Gray, 1837: 375; Hanley, 1843 (in 1842–1856): 34; Hupé in Gay, 1854: 348; Philippi, 1893: 11; Signorelli and Pastorino, 2011: 52; Soot-Ryen, 1959: 65.

Mulinia byronensis **Gray, 1837 (Fig. 2.4c–h)**

Shell oval to trigonal, equilateral, compressed; dorsal margins curved; line from umbos to posterior end very clear; umbos prosogyrous, not inflated; ventral margin curved; external surface smooth with irregular growth lines, brownish periostracum; internally white; internal ligament placed in a trigonal chondrophore ventrally developed; right hinge with two anterior and two posterior lateral teeth, strong, each with one rounded cusp; two cardinal teeth on anterior side of chondrophore, posterior cardinal vertically oriented; left valve with one anterior and one posterior lateral tooth, anterior cardinal tooth oblique in relation to dorsal margin; ultrastructure of the shell with two layers, internal with cross lamellar complex and external with simple cross lamellar structure; V-shaped pallial sinus of one-third of shell length.

TYPE LOCALITY: "South America".
TYPE MATERIAL: NHMUK 196324, 5 syntypes.
GEOGRAPHIC DISTRIBUTION: Peru to Ushuaia, Argentina.
STRATIGRAPHIC DISTRIBUTION: Recent.
SYNONYMS: *Cumingia clerii* Adams, 1850: 24; *Mactra byroni*, *Mactra byroneana* Weinkauff, 1884: 101 (in Küster et al., 1841–1884), unjustified emendations; *Mactra jousseaumei* Mabille and Rochebrune, 1889: 106 (in Rochebrune and Mabille, 1889); *Mactra pencana* Philippi, 1893: 8; *Mactra lotensis* Philippi, 1893: 10; *Mactra magellanica* Philippi, 1893: 10.

REMARKS: the nominal species *Mulinia byroni* or *byroneana* proposed by Weinkauff (1884, in Küster et al., 1841–1884) must be considered unjustified emendations. After type material examination, *Cumingia clerii, Mactra jousseaumei, Mactra pencana, Mactra lotensis* and *Mactra magellanica* are confirmed as synonyms (Signorelli and Pastorino, 2011). However, the shell shape variation observed *Mulinia byronensis* and *M. edulis* suggest a genetic comparison to properly define both taxa.

RELEVANT LITERATURE

Ecological and Environmental studies: González et al., 2002; Güller and Zelaya, 2015; Linse, 1999.

Taxonomy: Coan and Kabat, 2017: 90–92; d'Orbigny, 1846 (in 1834–1847): 511; Gray, 1837: 376; Huber, 2010: 447; Lamy, 1918: 332; Reeve, 1854a: sp. 77; Signorelli and Pastorino, 2011: 53; Soot-Ryen, 1959: 65–66; Zelaya, 2016: 253.

Mulinia cleryana (d'Orbigny, 1846) (Fig. 2.5a–d)

Shell trigonal, inflated, inequilateral, umbos placed about half shell length, inflated and prosogyrate; maximum shell length measured 36 mm; external surface smooth without concentric ridges; postero-dorsal area defined by a conspicuous keel-like carina; antero-dorsal area wide, but not defined by a lunule; anterior and posterior ends low and well defined; ventral margin sinuous; right hinge with two anterior and two posterior lateral teeth (AI, AIII, PI, and PIII) strong but short, being the ventral more elongated, two divergent and unfused cardinal (3a and 3b) teeth, the 3b vertically oriented; left hinge with the usual V-shaped cardinal tooth (2a-b), flanked by an accessory lamella (4b), one anterior and one posterior lateral tooth (AII and PII) complete the hinge; anterior adductor muscle scars semielliptical, posterior oval; pallial sinus low and V-shaped; shell ultrastructure with two layers as previous species.

TYPE LOCALITY: Rio de Janeiro, Brazil.

TYPE MATERIAL: NHMUK 1854.12.4.670, three syntypes.

GEOGRAPHIC DISTRIBUTION: Gulf of Mexico to Rio Grande do Sul, Brazil.

STRATIGRAPHIC DISTRIBUTION: Middle Miocene—Recent.

SYNONYMS: *Mactra guadelupensis* Récluz, 1852: 249; *Mactra carinulata* Reeve, 1854a: sp. 38; *Mulinea portoricensis* Shuttleworth, 1856: 174; *Gnathodon cantrainei* Sowerby in Reeve, 1873: sp. 2 and *Mulinia branneri* Dall, 1901: 145.

REMARKS: After type material examination, *Mactra guadelupensis* (MNHN-IM-2000-5017), *Mulinea portoricensis* (NMBE 501777); *Mactra carinulata* (NHMUK 1854.67.10.2.55), *Gnathodon cantrainei* (types not found) and *Mulinia branneri* (USNM 107829) are confirmed as junior synonyms of *Mulinia cleryana*. Although the examination of type material confirms the synonymy list, two morphotypes are observed, the Caribbean and the Brazilian. Additional studies based on morphometry and DNA analysis will confirm this appreciation.

RELEVANT LITERATURE

Ecological and Environmental studies: De Castro and Dos Santos, 1989: 103; Salvador et al., 1998.

Fig. 2.5 (**a–d**) *Mulinia cleryana* (d'Orbigny, 1846), NHMUK 1854.12.4.670, syntypes; (**e–h**) *Mulinia edulis* (King, 1832), NHMUK 1837.1.12.1168, syntypes. Scale bar (**a–d**) 1 cm, (**e–h**) 2 cm

Paleontology: Altena, 1968: 159, 1969: 17, 1971: 51; Weisbord, 1964: 382; Woodring, 1982: 650.

Taxonomy: Adams and Adams, 1856 in 1853–1858: 380; Conrad, 1868: 32; d'Orbigny, 1846 (in 1834–1847): 510; Dall, 1894c: 27, 1901: 145; Gundlach, 1883a: 58; Krebs, 1864: 105; Lamy, 1918: 341; Palacio et al., 2010: 407; Récluz, 1852: 249; Reeve, 1854a: sp. 38; Rios, 1975: 236, 1994: 266, 2009: 537; Robinson and Montoya, 1987: 396; Shuttleworth, 1856: 174; Signorelli and Pastorino, 2012b: 42; Sowerby G. B. III in Reeve, 1873: sp. 2; Tallarico et al., 2014: 101; von Cosel, 1973: 52; Warmke and Abbott, 1961: 204; Weinkauff, 1879: 33 (in Küster et al., 1841–1884).

Mulinia edulis (King, 1832) (Fig. 2.5e–h)

Shell trigonal, subcircular, strong; umbos slightly inflated, prosogyrous; external surface smooth; posterior dorsal margin rounded from umbos to posterior end; anterior dorsal margin straight; ventral edge curved without sharp angles; with reddish brown periostracum; right valve hinge with small, short lateral teeth; cardinal teeth unfused, left valve with V-shaped cardinal teeth composed of two simple, divergent cardinals; one anterior and one posterior lateral tooth present; ultrastructure of shell as in other species analyzed; pallial sinus separated from ventral edge, U-shaped, horizontal, approximately half of shell length; siphons fused; apertural area of incurrent characterized by the presence of tentacles in two rings; internal ring with two sizes of tentacles, external ring with only one; excurrent siphon with subequal tentacles; demibranchs posteriorly fused, type C (1) of Atkins (1937); inner demibranch with well-defined food groove over the distal edge; outer and inner demibranchs similar in size and shape; labial palps elongated, trigonal, with internal surfaces plicate facilitating food transport to mouth.

TYPE LOCALITY: Puerto El Hambre, Magellanic Strait, Chile.
TYPE MATERIAL: NHMUK 1837.1.12.1168–9 and 196325, three syntypes.
GEOGRAPHIC DISTRIBUTION: From Guayaquil, Ecuador to Ushuaia, Argentina.
STRATIGRAPHIC DISTRIBUTION: Pliocene—Recent.
SYNONYMS: ?*Mactra cojimbensis* Anton, 1837: 282; *Mulinia typica* Gray, 1837: 375; *Mactra rodatzi* Dunker, 1850: 30; *Mactra marcida* Gould, 1850: 216; *Mactra cuneola* Gould, 1850: 216; *Mactra gabbi* Tryon, 1870: 170; ?*Mulinia antarctica* Paetel, 1883: 201, "ex Dunker 1850"; *Mactra coquimbana* Philippi, 1887a: 254, 263 [1887b: 244, 353].
REMARKS: The examination of type specimens of *Mulinia typica* (NHMUK 20190677), *Mactra cuneola* (USNM 5891) and *Mactra marcida* (USNM 5892) both collected by the U. S. Exploring Expedition from Orange Bay, *Mactra gabbi* (ANSP 51392, syntypes) and *Mactra coquimbana* (MNHNS 212), confirmed that all nominal species must be considered as junior synonyms of *M. edulis*. Moreover, *Mactra rodatzi* Dunker, 1850 was confirmed in the genus *Mulinia* by Huber (2010), who suggested that the type locality is wrong. In addition, the observed characters from the original illustration distinct it from the other large mactrids known from East Africa. All these arguments suggest that Dunker species would be a junior synonym of *Mulinia edulis* as suggested by Huber (2010). However, the types were not found as deposited in the collection of the Museun für Naturfunde to confirm it. Finally, the types of *Mactra cojimbensis*, described from Cojimbo, probably Coquimbo, Chile (Anton, 1837), and *Mulinia antarctica*, mentioned from Chiloe Island, Chile (Paetel, 1883), were not found.

Relevant literature

Anatomy and Morphology: Coelho et al., 2016; Fischer, R. 1915; Signorelli and Pastorino, 2011; 55–58.

Ecological and Environmental studies: Garrido et al., 2012; González et al., 2002; Navarro et al., 2003; Velasco and Navarro, 2002.

Fisheries: Abarca et al., 2012, 2018; Avellanal et al., 2002; Díaz and Campos, 2014; Oliva et al., 2013; Vivanco et al., 2014.

Paleontology: Aguirre, 2003; Aguirre et al., 2008; Martinelli et al., 2017.

Taxonomy: Adams and Adams, 1856: 379; Alamo and Valdivieso, 1997: 128; Aldea and Valdovinos, 2005: 396; Bernard, 1983: 40; Carcelles and Williamson, 1951: 343; Carcelles, 1950: 81; Conrad, 1868: 31; d'Orbigny, 1846 (in 1834–1847): 510; Dall, 1909: 274; Gould, 1850: 216; Gray, 1837: 376; Hanley, 1843 (in 1842–1856): 35; Huber, 2010: 447; Hupé in Gay, 1854: 348; Paredes et al., 2016: 141; Philippi, 1893; Ramírez et al., 2003: 268; Reeve, 1854a: sp. 32; Reid and Osorio, 2000: 137; Rochebrune and Mabille, 1889: 106; Signorelli and Pastorino, 2011; 55; Soot-Ryen, 1959: 66.

Mulinia exalbida Gray, 1837 (Fig. 2.6a–d)

Shell oval, subcircular, inequilateral, strong, large, length to 73 mm; umbos prosogyrous, located approximately at midpoint of shell length, not in contact; with irregular growth lines very well defined, protruded from external surface; periostracum thin, yellowish, commonly absent in the umbonal area; posterior dorsal margin rounded, convex behind umbos, descending gently to posterior end and merging into the ventral rounded margin with a defined angle; lunule and escutcheon not delimited by any ornamentation; internally white; hinge formula as typical for Mactridae, right valve lateral teeth short and very close to cardinals; anterior ventral lateral teeth with sub-nodular morphology not observed in more elongated dorsal lateral teeth; cardinal teeth unfused, on anterior side of ventrally well-developed chondrophore; two posterior lateral teeth similar in size and shape; left valve lateral teeth short but deep; cardinal teeth fused, V-shaped, flanked by accessory lamellae; shell ultrastructure coinciding with that of previous species analyzed; adductor muscle scars circular, conspicuous, similar in size and shape; pallial sinus short, U-shaped, rounded and high.

Type locality: "South America" only mentioned in the original description.

Type material: NHMUK 1837.12.1.3688, holotype and one paratype.

Geographic distribution: Concepción, Chile, to Magellanic strait.

Stratigraphic distribution: Recent.

Synonyms: *Mactra cibaria* Philippi, 1893: 9; *Mactra calbucana* Philippi, 1893: 9; *Mactra jonasi* Philippi, 1893: 11.

Remarks: *Mulinia exalbida* has been neglected in the twentieth century South American literature (Carcelles, 1950; Carcelles and Williamson, 1951) probably because of its indefinite original type locality. However, it is a valid species that

Fig. 2.6 (**a–d**) *Mulinia exalbida* Gray, 1837, NHMUK 1837.12.1.3688, holotype; (**e–h**) *Mulinia lateralis* (Say, 1822), ANSP 52663, syntype. Scale bar (**a–d**) 2 cm, (**e–h**) 1 cm

inhabits the Magellanic Province (Signorelli and Pastorino, 2011). The examination of the types of *Mactra calbucana* (MNHNS 232), *Mactra cibaria* (MNHNS 206) and *Mactra jonasi* (MNHNS 229) confirmed all species as junior synonyms of *Mulinia exalbida*.

Relevant literature

Taxonomy: Adams and Adams, 1856: 380; Bernard, 1983; 40; Coan and Kabat, 2017: 90–92; Conrad, 1868: 31; Gray, 1837:376; Hanley, 1843 (in 1842–1856): 35; Philippi, 1893; Reeve, 1854a: sp. 78; Signorelli and Pastorino, 2011: 58.

Mulinia lateralis (Say, 1822) (Fig. 2.6e–h)

Shell rounded, trigonal, small to medium size, length to 15 mm; umbos subcentral prominent; radial ridge from umbos to posterior end solid and visible; external surface smooth but with very minute transversal wrinkles; anterior end rounded defined by a well-marked angle, posterior end acute; internally white, hinge plate small, not inclined with cardinal and lateral teeth closer, chondrophore triangular and slightly ventrally projected.

TYPE LOCALITY: South Carolina, United States.

TYPE MATERIAL: ANSP 52663, four syntypes.

GEOGRAPHIC DISTRIBUTION: north-western Atlantic from Canada to Central America.

STRATIGRAPHIC DISTRIBUTION: Early Pliocene—Recent.

SYNONYMS: *Mactra rostrata* Philippi, 1849: 152–153 (*non* Spengler, 1802).

REMARKS: After the examination of type material, the only junior synonym of this species is *Mactra rostrata* (MNHNS 213, holotype), described from Galveston, Texas, United States. On the other hand, *Mactra corbuloides* must be excluded from the synonymy due to spisuloid dentition (Smith, 1914; Tomlin, 1924; Huber, 2010).

RELEVANT LITERATURE

Ecological and Environmental studies: Craeymeersch et al., 2019; Cripe, 2006; Moss, 1989; Olszewski and Kaufman, 2015.

Paleontology: Bright et al., 2018; Daley, 2002; Klompmaker, 2009; Levinton, 1970; Olsson, 1922: 265; Pekar et al., 2004; Spencer and Campbell, 1987: 47; Staff et al., 1986; Woodring, 1982: 650.

Taxonomy: Abbott, 1954: 449; Brunel et al., 1998: 129; Camp et al., 1998: 12; Cauquoin, 1967: 225; Dall, 1894c: 27, 1898a: 901–902; Gould, 1841: 54; Lindell et al., 2006: 750; Mikkelsen and Bieler, 2007: 372; Morris, 1951: 97; Reguero and Raz-Guzmán, 2018: 36; Turgeon et al., 2009: 730; Zim and Ingle, 1955: 156.

Mulinia levicardo (E. A. Smith, 1881) (Fig. 2.7a–d)

Shell oval to subcircular, subequilateral, inflated, with small umbos weakly inflated; external surface smooth with weak concentric ornamentation forming irregular growth lines more visible along the ventral area; anterior and posterior ends rounded, without sharp angles, ventral margin curved; lunule and escutcheon poorly developed, not defined by lines; internally white; right valve with two strong anterior and posterior lateral teeth with rounded cusps, ventral anterior and posterior more distal from cardinals; cardinal teeth unfused, posterior one vertically oriented, anterior one oblique; left valve with V-shaped cardinal tooth flanked on posterior side by an accessory lamella, lateral teeth elongated with rounded cusp; chondrophore small, ventrally poorly developed; pallial sinus short and U-shaped.

Fig. 2.7 (**a–d**) *Mulinia levicardo* (E. A. Smith, 1881), NHMUK 1879.10.15.67, syntype; (**e–h**) *Mulinia modesta* Carpenter, 1864, USNM 3943, syntypes. Scale bar (**a–h**) 2 cm

Type locality: Cockle Cove, in Magellanic Strait area, at 13 m depth.
Type material: NHMUK 1879.10.15.67, syntype.
Geographic distribution: Magellanic Strait area.
Stratigraphic distribution: Recent.
Synonyms: None.

Remarks: This species is rare, very scarce in the examined malacological collections. It was collected during the trip of the H.M.S. "Alert". Signorelli and Pastorino (2011) revised the type material and recognized it as a valid species. No synonyms were found in the literature.

Relevant literature

Ecological and Environmental studies: Aldea and Valdovinos, 2005; González et al., 2002; Linse, 1999.

Taxonomy: Carcelles, 1950: 80; Carcelles and Williamson, 1951: 343; Rochebrune and Mabille, 1889: 106; Signorelli and Pastorino, 2011: 60; Smith E. A., 1881: 39.

Mulinia modesta (Carpenter, 1864) (Fig. 2.7e–h)

Shell sub-trigonal, inflated to highly inflated, small to medium size, length to 70 mm, equilateral, umbos prominent, inflated, placed at half of shell length; external surface smooth; postero-dorsal area slightly defined by a hardly angulated ridge on which the periostracum forms a line, anterior end rounded, posterior end acute; periostracum brownish, thin, dehiscent; internally white, hinge plate small with an exclusively internal ligament placed into a well-developed and subcircular chondrophore; pallial sinus V-shaped, moderately short and narrow.

Type locality: Guaymas, Sonora, Mexico.

Type material: USNM 3943, syntypes, four specimens.

Geographic distribution: West coast of United States from Baja California to Teacapán, Sinaloa, Mexico.

Stratigraphic distribution: Pleistocene—Recent.

Synonyms: *Mactra coloradoensis* Dall, 1894b and *Mulinia coloradoensis acuta* Dall, 1894b.

Remarks: The authority of *Mulinia modesta* must be attributed to Carpenter, 1864 who mentioned this name, based on Sloat ms. Not to Dall (1894b: 5) who cited *Mulinia modesta*, based on the same specimen collected by Mr. Sloat. After type material examination of *Mactra coloradoensis* (USNM 36570, syntypes) and *Mulinia coloradoensis acuta* (USNM 117506, syntypes), both species are confirmed as junior synonyms of *M. modesta*. There is an extensive literature of this species mainly mentioned as *Mulinia coloradoensis*, focused on ecology and paleoecology of Colorado River basin.

Relevant literature

Ecological and Environmental studies: Avila-Serrano et al., 2006; Brusca et al., 2017; Cintra-Buenrostro et al., 2012; Dietl and Smith, 2017; Flessa, K. W. and Rodriguez, C. A. 1999; Kowalewski et al., 2000; Smith and Dietl, 2019; Smith et al., 2018 (mostly as *M. coloradoensis*).

Taxonomy: Coan and Valentich-Scott, 2012: 552; Flessa, 2002; Flessa and Téllez-Duarte, 2001; Gemmell et al., 1987: 45; Grant and Gale, 1931: 401; Keen, 1971: 207; Kowalewski and Flessa, 1995; Kowalewski et al., 2000; Rodriguez et al., 2001.

Mulinia pallida (Broderip and G. B. Sowerby I, 1829) (Fig. 2.8a–d)

Shell sub-trigonal, subequilateral, moderately elongated at the posterior end; thin, inflated; medium size to large, length to 93 mm; umbos broad, inflated, anterior end rounded and sharply; posterior end low and pointed; posterior slope demarked by a shelly ridge forming a sharp angle; external surface smooth with irregular growth

Fig. 2.8 (**a–d**) *Mulinia pallida* (Broderip and G. B. Sowerby I, 1829), NHMUK 1858.5.12.49, holotype; (**e–h**) *Rangia cuneata* (Sowerby I, 1832, in 1821–1834), NHMUK 1970097, syntype. Scale bar 2 cm

lines; yellowish periostracum, usually absent in the umbonal area; internally white, hinge plate elongated, with lateral teeth slightly distant from cardinal area; chondrophore moderately projected; ligament exclusively internal; pallial sinus moderately short and narrow.

TYPE LOCALITY: San Blas, Tepic, Mexico.

TYPE MATERIAL: NHMUK 1858.5.12.49, holotype.

GEOGRAPHIC DISTRIBUTION: From west coast of North America, Gulf of California, Mexico to Peru.

STRATIGRAPHIC DISTRIBUTION: Miocene—Recent.

SYNONYMS: *Mulinia donaciformis* Gray, 1837: 376; *Mactra angulata* Reeve, 1854a: sp. 34; *Mactra goniata* Deshayes, 1855a: 70; *Mactra lacinata* Carpenter, 1856: 161; *Mactra (Mulinia) bistrigata* Mörch, 1861: 182; *Mulinia bradleyi* Dall, 1894b: 6; *Corbula altirostris* Li, 1930: 263; *Mulinia camina* Pilsbry and Olsson, 1941: 74.

REMARKS: This species was largely named by the description of Recent and fossil specimens. Type material examination of the Recent *Mulinia donaciformis* (NHMUK 196326, syntype), *Mactra angulata* (NHMUK 196660, syntype); *Mactra goniata* (described from California, type not found), *Mactra lacinata* (NHMUK 19621117, holotype); *Mactra (Mulinia) bistrigata* (ZMUC-BIV-001549), *Mulinia bradleyi* (USNM 101998, syntypes) and *Corbula altirostris* (AMNH 268092, holotype) confirmed all nominal species as junior synonyms of *M. pallida*. Finally, the fossil species *Mulinia camina* (ANSP 13700, holotype, ANSP 78997, paratype), described from Jama Formation—Punta Cabuyal, Ecuador is the same.

RELEVANT LITERATURE

Ecological and Environmental studies: Esqueda-González et al., 2014: 52.

Paleontology: Cantalamesa et al., 2005: 5; Cruz, 1983: 752; Olsson, 1964: 64; Pilsbry and Olsson, 1941: 73; Woodring et al., 1946: 88; Woodring, 1982: 651.

Taxonomy: Cauquoin, 1969: 1028–1029; Coan and Valentich-Scott, 2012: 552–554; Dall, 1894d: 41; Grant and Gale, 1931: 400–401; Hertlein and Strong, 1950: 233; Keen, 1971: 207; Lamy, 1918: 335; López de Mesa and Cantera, 2015: 8; Olsson, 1961: 330–331; Paredes et al., 2016: 141.

Genus **Rangia** *des Moulins, 1832* **(non** *Férussac, 1837 in Férussac et al., 1819–1851,* **non** *Agassiz, 1860)*

[= *Gnathodon* Sowerby, 1832, in 1821–1834: 36 (*non* Oken, 1816: 140); *Colombia* (Blainville MS) Rang, 1834: 217 (*nomen nudum*); *Perissodon* Conrad, 1862: 573 (type, *Mactra clathrodonta* Conrad, 1833: 340 by subsequent designation, Dall, 1894e: 99)]

Shell trigonal, posterior end elongated, lateral teeth strongly elongated, the posterior longer; ligament exclusively internal.

TYPE SPECIES: *Rangia cyrenoides* des Moulins, 1832: 57 (= *Gnathodon cuneatus* Sowerby I, 1832, in 1821–1834: 36), by monotypy.

GEOGRAPHIC DISTRIBUTION: East and West North America, West South America, Europe. The genus *Rangia* was registerd from Upper Cretaceous deposits (Richards, 1962a), but mainly reported from Pleistocene deposits of southern United States (Cooke 1936; Richards, 1962b; Coch, 1971; Miller 1983; Portell et al., 1992; Hershler and Jayko, 2009).

REMARKS: This is a small genus of mactrid bivalves, currently distributed in estuarine waters of the southern Atlantic coast of United States to the Gulf of Mexico, and Gulf of California in the Pacific (Dall, 1894e; Morrison, 1959; Keen, 1971; Abbott, 1974; Wakida-Kusunoki and MacKenzie, 2004; Coan and Valentich-Scott, 2012). It is represented by four valid species: *Rangia (Rangia) cuneata* (Sowerby I, 1832, in 1821–1834), *R. (Rangianella) flexuosa* (Conrad, 1840), *R. (Rangianella) mendica* (Gould, 1851) and *Rangia (Rangianella) pontchartrainensis* (Morrison, 1965) (new combination).

Subgenus Rangia *des Moulins, 1832*

Rangia (Rangia) cuneata (Sowerby I, 1832) (Fig. 2.8e–h)

Shell obliquely ovate, thick and heavy; medium size, length to 65 mm; umbos prosogyrate, inflated, not in touch; external surface smooth, with irregular growth lines covered with a strong gray-brown periostracum; antero-dorsal margin short, postero-dorsal margin longer and slightly convex to straight; anterior end high and rounded, posterior end low and acute; interiorly glossy, white; hinge plate strong with anterior lateral teeth hooked; posterior lateral very elongated; chondrophore, postero-ventrally oriented; pallial sinus short but well distinct.

TYPE LOCALITY: New Orleans, United States.

TYPE MATERIAL: NHMUK 1970097, one syntype.

GEOGRAPHIC DISTRIBUTION: From Chesapeake Bay, United States to Gulf of Mexico.

STRATIGRAPHIC DISTRIBUTION: Upper Pliocene—Recent.

SYNONYMS: *Rangia cyrenoides* Des Moulins, 1832: 57; *Gnathodon nasutus* Dall, 1894e: 98.

REMARKS: *Rangia cuneata* was widely mentioned in the literature. Several publications focused on sedimentology, ecology and fisheries in the Gulf of Mexico were carried out. In the Relevant Literature section, some of them are mentioned. Recently, it was reported as introduced in European waters (Verween et al., 2006; Katsanevakis et al., 2012; Wiese et al., 2016). A lot of Des Moulins with three specimens deposited at the Muséum d'Histoire naturelle de Bordeaux (MHNBx 2009.21540.0) was examined. The lot contains three specimens, but one (MHNBx 2009.21540.3, without type status, length 73 mm, height 60 mm) is larger than the maximum size indicated by Des Moulins. However, the other two specimens (MHNBx 2009.21540.1 syntype, length 45 mm, height 41 mm and MHNBx 2009.21540.2 syntype, length 30 mm, height 27 mm) fit with the reported size in the description. An old label of Des Moulins confirmed the type status for two out of the three specimens. The examination of related type material confirmed *Rangia cyrenoides*, collected on Lake Pontchartrain, United States, as a junior synonym of *R. cuneata*. The examination of the holotype of *Gnathodon nasutus* (USNM 106988) confirmed it as a junior synonym of *R. cuneata*. Finally, *Gnathodon grayi* Conrad, 1838 (unnecessary new name for *Mactra clathrodonta* Conrad, 1833: 340, ANSP 16242, syntypes), recorded from Upper Miocene of Yorktown, and *Gnathodon minor* Conrad, 1841 (ANSP 16247, syntypes), from Pliocene of North Carolina, were mentioned as junior synonyms of *R. (R.) cuneata* by Tuomey and Holmes (1857) and Holmes (1860). However, the Pleistocene specimens reported by Tuomey and Holmes (1857) and Holmes (1860) were considered by Dall (1894e) as misidentifications of Conrad's species but definitely as synonyms of *R. cuneata*. More recently, Spencer and Campbell (1987) reported morphological differences between the Upper Pliocene—Recent *R. cuneata* and the Upper Miocene *R. clathrodonta*.

Relevant literature

Anatomy and Morphology: Fritz et al., 1990; Rogers and García-Cubas y Gutiérrez, 1981.

Ecological and Environmental studies: Andrus et al., 2008; Bedford and Anderson, 1972; Chanley, 1965; Cooper, 1981; Gallagher and Wells, 1969; Godwin, 1968; Hopkins, and Andrews, 1970; Verween et al., 2006.

Fisheries: Cain, 1975; Carpenter, 2002: 61; Hopkins et al., 1974; Jovanovich and Marion, 1989; Sundberg and Kennedy, 1992, 1993; Swingle and Bland, 1974; Wakida-Kusunoki and MacKenzie, 2004 (and literature therein).

Paleontology: Conrad, 1833: 340, 1838: 23, 1845: 69; Conrad in Hodge, 1841: 347; Cooke 1936: 152; DuBar 1962: 65; Fairbanks, 1963; Foltz et al., 1995; Holmes, 1860: 41; Metcalf, 1980; Petuch and Roberts 2007: 176–179; Richards, 1962b: 69; Spencer and Campbell 1987: 47; Tenore et al., 1968; Tuomey and Holmes, 1857: 99.

Taxonomy: Abbott, 1974; Camp et al., 1998: 12; Dall, 1894e: 97; Hill, 1983; Huber, 2010: 447; Lamy, 1918: 344; Turgeon et al., 2009: 730.

Subgenus Rangianella *Conrad, 1868*

Small, nearly equilateral, lateral teeth shorter than in *Rangia* s.s., straight, subequal, faintly rugose; pallial sinus inconspicuous or obsolete. Estuarine to marine.

Type species: *Gnathodon trigonum* Petit de la Saussaye, 1853a (= *Mactra mendica* Gould, 1851), by monotypy.

Geographic distribution: West Central and North America.

Rangia (Rangianella) flexuosa (Conrad, 1840) (Fig. 2.9a, b)

Shell ovate, thick, small to medium size, length to 37 mm, nearly equilateral, umbos prominent, posterior slope with a keel-like process, large lunule; external surface smooth with irregular growth lines; internally white to greyish; hinge plate with very short lateral teeth, pallial sinus obsolete.

Type locality: Florida, United States.

Type material: ANSP 52653, syntype, a single right valve.

Geographic distribution: Gulf of Mexico from Florida, United States to Veracruz, Mexico.

Stratigraphic distribution: Recent.

Synonyms: *Gnathodon rostratum* Petit de la Saussaye, 1853a: 84; *Gnathodon flexuosa petitianus* Dall, 1894e: 103.

Remarks: *R. (Rangianella) flexuosa* is an Atlantic species commonly recorded in the Gulf of Mexico. It can be distinguished from *R. cuneata* by the presence of shorter lateral teeth and more elongated posterior end. Type material examination of *Gnathodon rostratum* (MNHN-IM-2000-34897, syntype) and *Gnathodon flexuosa petitianus* (USNM 57668A, holotype) confirmed both species as junior synonyms of *R. (R.) flexuosa*.

Relevant literature

Ecological and Environmental studies: Foltz et al., 1995.
Fisheries: Wakida-Kusunoki and MacKenzie, 2004 (and literature therein).
Paleontology: Simkins et al., 2012; Troiani et al., 2011.
Taxonomy: Abbott, 1974; Dall, 1894c: 27, 1894e; Lamy, 1918: 346; Prime, 1860; Reeve, 1873: sp. 5; Reguero and Raz-Guzmán, 2018: 36; Turgeon et al., 2009: 730.

Rangia (Rangianella) mendica (Gould, 1851) (Fig. 2.9c–h)

Shell trigonal, small to medium size, length to 30 mm, inflated, thick, elongated posteriorly; anterior end rounded, posterior end pointed and low; external surface smooth, with commarginal fine striae; internally white, hinge plate with small cardinal teeth and short but strong lateral teeth; internal ligament in small, deep rounded chondrophore; pallial sinus very shallow.

Type locality: Mazatlán, Sinaloa state, Mexico.

Type material: Not found, probably lost. Neotype herein designated USNM 103899, one complete specimen from Mazatlán, Mexico.

Geographic distribution: Pacific coast of Mexico, Gulf of Venezuela.

Stratigraphic distribution: Recent.

Synonyms: *Gnathodon trigonum* Petit de la Saussaye, 1853a: 84, 1853b: 166; *G. truncatum* Petit de la Saussaye, 1853b: ii (*nomen nudum* in figured caption).

Fig. 2.9 (**a**, **b**) *Rangia (Rangianella) flexuosa* (Conrad, 1840), ANSP 52653, syntype; (**c**–**h**) *Rangia (Rangianella) mendica* (Gould, 1851), (**c**–**f**): USNM 103899, Neotype from Mazatlan, Mexico (herein designed); (**g**, **h**) SBMNH 352588, Nayarit, Mexico; (**i**–**l**) *Rangia (Rangianella) pontchartrainensis* (Morrison, 1965), USNM 635643, holotype. Scale bar (**a**–**h**) 1 cm, (**i**–**l**) 5 mm

Remarks: *Rangia (Rangianella) mendica* is the only Recent species of the genus living in the west coast of North America. However, it was recently reported as introduced in Golfo de Paria and Orinoco delta (Pérez et al., 2007). Type material examination of *Gnathodon trigonum* (MNHN-IM-2000-34896, syntype) confirmed it as a junior synonym of *R. (R.) mendica.*

Relevant literature

Ecological and Environmental studies: Pérez et al., 2007.
Taxonomy: Coan and Valentich-Scott, 2012: 554; Dall, 1894e: 102; Félix-Pico et al., 2011: 214; Keen, 1971: 209; Poutiers, 1995: 145; Prime, 1860: 348.

Rangia (Rangianella) pontchartrainensis (Morrison, 1965) (new combination) (Fig. 2.9i–l)

Shell small, trigonal, equivalve, umbos low, smooth, posterior ridge low, bi-angulate; anterior end rounded; external surface porcelaneous; with thin periostracum more evident on posterior ridge; hinge plate typical of the group, with V-shaped cardinal tooth in the left valve and lateral teeth short and smooth, close to the cardinals; anterior adductor muscle scar smaller; ligament exclusively internal placed on ventral surface of chondrophore; pallial sinus shallow.

Type locality: collected from "Middle Ground" in the eastern part of Lake Pontchartrain, Louisiana, United States.

Type material: USNM 635643, holotype, and approximately a thousand paratypes USNM 635644 collected on November 5, 1959.

Geographic distribution: Brackish waters of southeastern United States, from Lake Pontchartrain and Borgne, Louisiana, Mississippi from Heron, St. Louis and Back bays, and from Davis Bayou.

Stratigraphic distribution: Recent.

Synonyms: None.

Remarks: This brackish species, from Lakes Pontchartrain and Borgne, and Mississippi river, was described as a *Mulinia*. In the original description, the author compared the type specimens with other living mactrids from the same area. The analysis of types suggests morphological affinities with *Rangia (Rangianella)*. Hinge plate morphology, elongate posterior end and very short pallial sinus are similar to other species of the subgenus.

Relevant literature

Ecological and Environmental studies: Brown et al., 2013.
Taxonomy: Junot et al. 1983: 250; Morrison, 1959: 26, 1965: 222–223; Sikora et al., 1981; Sikora and Sikora, 1982.

Chapter 3
The Subfamily Mactrinae (Mollusca: Bivalvia), Part 2

Abstract In this chapter the species included in the genera *Simomactra*, *Spisula*, *Standella*, *Trinitasia* and *Tumbeziconcha* are redescribed. The genus *Simomactra* Dall, 1894d includes three Pacific species: *S. dolabriformis* (Conrad, 1867), *S. falcata* (Gould, 1850) and *S. planulata* (Conrad, 1837). *Spisula* is another genus with several living species mainly distributed in temperate to cold waters in the north and south regions of the continent. This genus groups the species *S. fuegiensis* (E. A. Smith, 1905), *S. guidoi* (Signorelli and Scarabino, 2010), *S. isabelleana* (d'Orbigny, 1841), *S. marplatensis* (Doello Jurado in Carcelles, 1949), *S. petitii* (d'Orbigny, 1841), *S. williamsi* (Berry, 1960), *S.* (*Hemimactra*) *solidissima* (Dillwyn, 1817) and *S.* (*Hemimactra*) *raveneli* (Conrad, 1832). The genus *Standella* Gray, 1853 with eight species is another group well represented in American marine waters. It includes the species *S.* (*Standella*) *fragilis* (Gmelin, 1791), *S.* (*Standella*) *nasuta* (Gould, 1851), *S.* (*Micromactra*) *angusta* (Reeve, 1854a), *S.* (*Micromactra*) *californica* (Conrad, 1837), *S.* (*Micromactra*) *janeiroensis* (E. A. Smith, 1915), *S.* (*Micromactra*) *surinamensis* (Altena, 1968), *S.* (*Mactroderma*) *isthmica* (Pilsbry and Lowe, 1932) and *S.* (*Mactroderma*) *velata* (Philippi, 1849). The genus *Trinitasia* Maury, 1928 with *T. iheringi* (Dall, 1897) and the genus *Tumbeziconcha* Pilsbry and Olsson, 1935 with *T. goniocyma* (Pilsbry and Lowe, 1932) and *T. thracioides* (Adams and Reeve, 1850) complete the chapter.

Keywords *Simomactra* · *Spisula* · *Standella* · *Trinitasia* · *Tumbeziconcha*

Genus *Simomactra* Dall, 1894d

[= *Symmorphomactra* Dall, 1894d, p. 41 (type, *Mactra falcata* Gould, 1850, by monotypy]

Shell large, thick, posterior area defined by a line, like in *Standella*; with chondrophore moderately projected; external ligament separated from internal by a shelly ridge; pallial sinus U-shaped, anterior end elongated; hinge plate with lateral teeth pustules on the dorsal surface.

Type species: *Spisula dolabriformis* Conrad, 1867 by monotypy.

J. H. Signorelli, *The Superfamily Mactroidea (Mollusca:Bivalvia) in American Waters*, https://doi.org/10.1007/978-3-030-29097-9_3

Geographic distribution: Pacific coast of North America to Central America, South Africa, Indian Ocean.

Remarks: This genus is well represented in the west coast of North America and Central America by three species (Coan et al., 2000, Huber, 2010; Coan and Valentich-Scott, 2012). They are the type species *Simomactra dolabriformis* (Conrad, 1867), *S. falcata* (Gould, 1850) and *S. planulata* (Conrad, 1837). Two additional valid species, distributed from South Africa to the Indian Ocean are currently included in *Simomactra*. They are *S. capensis* (Sowerby III, 1892) and *S. chionia* (Tomlin, 1921). Huber (2010) mentioned morphological similarities with the type species *S. dolabriformis* to tentatively place both taxa in *Simomactra*. After the analysis of all related types, those similarities (shape, periostracum, pallial sinus and hinge plate) are not so evident. Genetic analysis is necessary to resolve this issue.

Simomactra dolabriformis *(Conrad, 1867) (Fig. 3.1a–d)*

Shell ovate to elongate, equilateral, slightly ventricose, compressed, thick; large, length to 94 mm; external surface smooth with a periostracum thin, silky, dark brown, much wrinkled on posterodorsal slope; anterior end longer, rounded at the end; posterior end gaped narrowly and obliquely truncated; ventral margin regularly rounded; posterodorsal area defined by a carina; internally white, hinge plate with short lateral teeth placed near the cardinals; chondrophore not ventrally projected; pallial sinus U-shaped, moderate in length, not reaching the umbo.

Type locality: Panama is mentioned in the original description. However, California is mentioned in the label of types.

Type material: ANSP 51411, syntype, 2 valves.

Geographic distribution: West coast of America from Bahía Magdalena, Gulf of California to Panama, recorded in Atacames, Esmeraldas, Ecuador.

Stratigraphic distribution: Pliocene—Recent.

Synonyms: None.

Remarks: No synonyms were found in the literature. Coan and Valentich-Scott (2012) mentioned that records of this species from California were based on misidentification of *S. planulata* or *S. falcata*. The thinner and longer cardinal teeth in the left valve separates *S. dolabriformis* from the other species of the genus.

Relevant literature

Paleontology: Hickman, 2015: 32; Packard, 1916: 279; Valentine, 1957.

Taxonomy: Bernard, 1983: 39; Cauquoin, 1969a: 1020–1021; Coan and Valentich-Scott, 2012: 554–555; Coan et al., 2000: 458; Dall, 1894a: 137, 1916a: 39; Gemmell et al., 1987: 44–45; Huber, 2010: 445; Keen, 1971: 202; Olsson, 1961: 323–324; Strong, 1925.

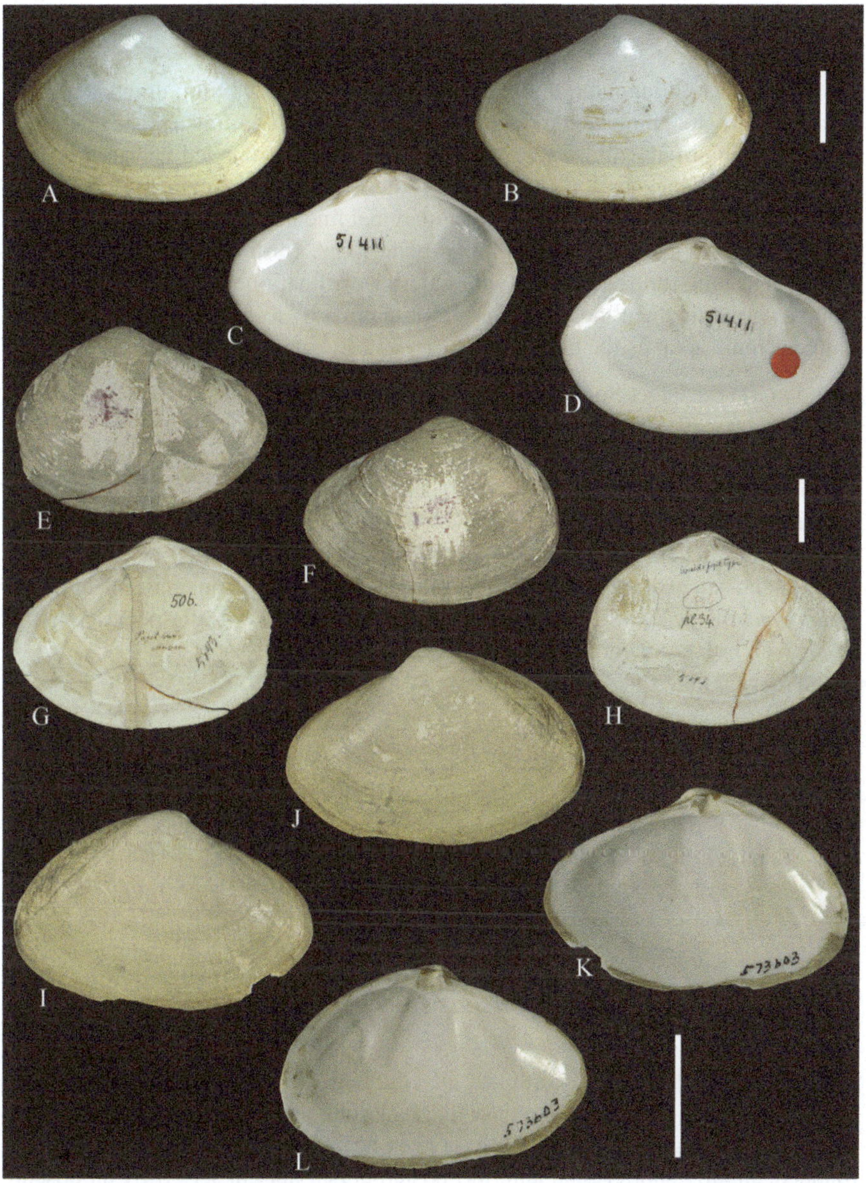

Fig. 3.1 (**a–d**) *Simomactra dolabriformis* (Conrad, 1867), ANSP 51411, syntypes; (**e–h**) *Simomactra falcata* (Gould, 1850), USNM 5893, holotype; (**i–l**) *Spisula strongi* Burch, 1945 (= *Simomactra planulata* (Conrad, 1837)), holotype USNM 573603. Scale bar (**a–d**) 1 cm, (**e–l**) 2 cm

Simomactra falcata *(Gould, 1850) (Fig. 3.1e–h)*

Shell large, ovate-triangular, thick, inequilateral, length to 90 mm; umbo acute, not inflated, nearly touching, anterodorsal margin straight, presenting a plane lanceolate space in front of the umbones; posterodorsal margin rounded; posterior slope defined by a change in sculpture only; posterior end rounded, narrowly gaped; external surface smooth, marked by commarginal fine lines; periostracum thin dark brownish, silky; internally white, hinge plate large, cardinal teeth very unequal, lateral teeth long, thin, left valve with anterior cardinal on same lamina as lateral tooth; external ligament not separated from internal by a shelly ridge; pallial sinus deep reaching as far as the umbo.

TYPE LOCALITY: Puget Sound, northwest coast of United States.

TYPE MATERIAL: USNM 5893, holotype.

GEOGRAPHIC DISTRIBUTION: Northwest coast of North America, from British Columbia, Canada to Baja California, United States.

STRATIGRAPHIC DISTRIBUTION: Pliocene—Recent.

SYNONYMS: *Anisodonta pellucida* Dall, 1916a: 30, *nom. nud.*; 1916b: 411.

REMARKS: Coan et al. (2000) mentioned that *A. pellucida* is a juvenile specimen of *Simomactra falcata*. The type material examination of Dall species (USNM 208475, holotype) confirmed the synonymy.

RELEVANT LITERATURE

Ecological and Environmental studies: Palm et al., 2012.

Paleontology: Grant and Gale, 1931: 396–397; Hertlein and Grant, 1972: 314; Hetherington et al., 2003: 1760; Packard, 1916: 297–298.

Taxonomy: Bernard, 1983: 40; Carpenter in Lea et al., 1860: 2; Coan and Valentich-Scott in Carlton, 2007: 848; Coan and Valentich-Scott, 1997: 17; Coan et al., 2000: 458; Dall, 1916a: 40; Huber, 2010: 445; Strong, 1925.

Simomactra planulata *(Conrad, 1837) (Fig. 3.1i–l)*

Shell sub-trigonal, compressed, subequilateral, large, length to 80 mm; umbo softly inflated; external surface smooth with a shining and brownish periostracum; anterodorsal margin straight, slightly convex near the anterior end; posterior end rounded, slightly shorter than anterior; posterodorsal margin set off by an obscure line; not posteriorly gaped; external ligament not separated from internal by a shelly ridge; internally white, hinge plate with lateral teeth short, thick, left valves with anterior lateral shorter than in *Simomactra falcata*, distant from the anterior cardinal tooth; pallial sinus shallow.

TYPE LOCALITY: Unknown.

TYPE MATERIAL: Not found, probably lost (sensu Keen, 1966).

Geographic distribution: West coast of North America from northern California, United States to Baja California, Mexico.

Stratigraphic distribution: Pleistocene—Recent.

Synonyms: *Spisula strongi* Burch, 1945: 50.

Remarks: After the examination of type material of *Spisula strongi* (USNM 573603, holotype), this nominal species is considered as a junior synonym of *S. planulata* as suggested Coan et al. (2000).

Relevant literature

Anatomy and Morphology: Kellogg, 1915: 648.

Paleontology: Grant and Gale, 1931: 397–398; Hertlein and Grant, 1972: 316–317; Packard, 1916: 293–294.

Taxonomy: Bernard, 1983: 40; Burch, 1945: 50; Carpenter in Lea et al., 1860: 2; Coan and Valentich-Scott in Carlton, 2007: 848; Coan and Valentich-Scott, 1997: 17; Coan et al., 2000: 458–459; Dall, 1916a: 40; Huber, 2010: 445; Strong, 1925.

Genus *Spisula* Gray, 1837

[= *Spisulina* Fischer, 1887: 1116 (type, *Mactra truncata* Montagu, 1808: 34); by monotypy]

Shell ovate, large, equilateral, smooth; posteriorly elongated, external ligament separated from the internal by a gap; lateral teeth serrate.

Type species: "*Mactra solida* Montagu" (= *Cardium solidum* Linnaeus, 1758: 1126), by subsequent designation (Gray, 1847)

Geographic distribution: Cosmopolitan.

Remarks: Eight valid species are herein included in the genus *Spisula*. The South American species included in *Mactra* by Signorelli and Pastorino (2011, 2012a, 2012b) show morphological characters that fit more adequately to *Spisula* as suggested by Huber (2010, 2015). However, six of these species need more analysis to arrive to a conclusion. Hinge plate morphology, shell shape, periostracum and pallial sinus suggest that *S. fuegiensis* (Smith, 1905), *S. guidoi* (Signorelli and Scarabino, 2010), *S. isabelleana* (d'Orbigny, 1841) and *S. marplatensis* (Doello-Jurado in Carcelles, 1949) are congeneric. *S. petitii* (d'Orbigny, 1841) and *S. williamsi* (Berry, 1960) show morphological characters that distinguish both taxa from the other American *Spisula*. This scenario suggests the presence of two new subgenera that must be genetically confirmed. At the moment, they are placed in *Spisula* (*Spisula*). The observations of Huber (2010) about the publication date of *S. petitii* and *S. isabelleana* is correct (both illustrated in pl. 77, 1841). On the other hand, there exist no hesitations about the taxonomic placement of *S. (Hemimactra) solidissima* (Dillwyn, 1817), the type species of the subgenus, and *S. (Hemimactra) raveneli* (Conrad, 1832) (See the Remarks of *Hemimactra*).

Spisula (Spisula) fuegiensis *(E. A. Smith, 1905) (Fig. 3.2a–f)*

Shell trigonal, inequilateral, strong, large, length to 65 mm; external surface smooth with commarginal lines close to the umbonal area; periostracum brownish and thin; anterior dorsal margin straight, posterior rounded; right hinge with two divergent cardinals (3a and 3b), unfused; anterior lateral teeth (AI and AIII), short and close to cardinals; posterior lateral teeth present (PI and PIII). V-shaped cardinal tooth in left valve (2a and 2b). One anterior and one posterior lateral tooth (AII and PII) running parallel to the dorsal shell margin; each elongated, fragile, with one cusp; external ligament rudimentary, posterior to umbos; internal ligament (resilium) large, strong, in chondrophore; muscle scar semicircular, vertically elongated, with dorsal and ventral ends pointed, close to dorsal edge; pallial sinus rounded, one-third of total shell length; ultrastructure of the shell with two layers, outer crossed lamellar of elongated crystallites and inner complex crossed lamellar; labial palps very elongated in relation to body size.

TYPE LOCALITY: San Sebastian beach, Tierra del Fuego, Argentina. Collected by Captain Crawshay in October 1904.

TYPE MATERIAL: NHMUK 1905.8.11.7-10, four syntypes. The dimensions of the specimen illustrated by E. A. Smith are length 51, 5 mm, height 37, 5 mm, and width 21 mm.

GEOGRAPHIC DISTRIBUTION: Golfo Nuevo in Argentina to Punta Arenas in Chile.

STRATIGRAPHIC DISTRIBUTION: Pleistocene—Recent.

SYNONYMS: *Mactra epidermia* Philippi, 1893: 9 (*non* Reeve, 1854a).

REMARKS: Based on hinge plate morphology, *Spisula fuegiensis* is herein recombined. The lateral teeth show pustules on the dorsal surface. However, a new subgenus could be possible to group this species with *S. guidoi*, *S. isabelleana* and *S. marplatensis* as suggested by Huber (2015: 836). Type material examination of *Mactra epidermia* Philippi (MNHNS 219, holotype) confirmed the synonymy suggested by Signorelli and Pastorino (2011).

RELEVANT LITERATURE

Anatomy and Morphology: Signorelli and Pastorino, 2011.
Paleontology: Richards and Craig, 1963.
Taxonomy: Carcelles, 1950: 81; Carcelles and Williamson, 1951: 343; Huber, 2010: 758; 2015: 836; Signorelli and Pastorino, 2011: 49–52; Zelaya, 2016: 252.

Spisula (Spisula) guidoi *(Signorelli and Scarabino, 2010) (Fig. 3.2g–j)*

Shell, elongate, elliptical, equilateral, moderate size to large, length to 45 mm; slightly prosogyrate umbo; external surface smooth, covered by brownish periostracum; posteroventral margin usually sinuated; a weakly defined ridge delineates a posterior area; lunule not well delineated; right hinge with two anterior (LAI and

Fig. 3.2 (**a–f**) *Spisula (Spisula) fuegiensis* E. A. Smith, 1905, NHMUK 1905.8.11.7-10, syntype, (**e**, **f**) detail of hinge plate; (**g–j**) *Spisula (Spisula) guidoi* (Signorelli and Scarabino, 2010), USNM 878150, paratype. Scale bar (**a–d**) 2 cm, (**e–j**) 1 cm

LAIII) and two posterior (LPI and LPIII) lateral teeth located close to cardinals, parallel and short, the ventral one being the strongest; two cardinal teeth, the posterior one (3b) dorsoventrally oriented, flanking the chondrophore; the anterior one (3a) anteriorly inclined; left hinge with inverted V-shaped tooth formed by two cardinal teeth (2a and 2b), the posterior one flanked by accessory lamellae (4b); two lateral teeth, one anterior and one posterior, each with only one cusp (LAII and LPII); ultrastructure of the shell composed of two aragonitic layers; outer cross-lamellar layer constituted by very thin lamellae; inner layer thinner, with complex cross-lamellar structure; Adductor muscles large; anterior slightly larger than posterior; foot laterally compressed, directed anteroventrally; pallial sinus short, rounded, U-shaped; labial palps triangular, very elongate, with inner surfaces plicate and the outer surfaces smooth; siphons Type C of Yonge (1948), covered completely with periostracum.

Type locality: 60 km E-NE of Cabo Polonio, Uruguay (34°10′S, 53°04′W).

Type material: Holotype: articulated valves MNHNM 15526; paratypes MNHNM 15527 (3 single valve, 2 articulated valves); MACN-IN 37507 (1 v); MNHN 21121 (1 v); FURG 30536 (1 v); USNM 878150 (1 v). All collected in the type locality. MNRJ 9261 (2 lv in alcohol), 23 km southeast of Chuy, Uruguay (33°52′00″S, 53°10′6″W; 22 m).

Geographic distribution: From Espiritu Santo, Brazil, to the Valdés Peninsula, Argentina.

Stratigraphic distribution: Pleistocene—Recent.

Synonyms: *Mactra patagonica* of authors (*non* d'Orbigny, 1846).

Remarks: This species was described by Signorelli and Scarabino (2010). It was confused with *Mactra patagonica* d'Orbigny, 1846 for many years by several authors. Morphological differences between the types of *Mactra patagonica* and *S. guidoi* were exposed after the comparison of type material of d'Orbigny species with the collected specimens of *S. guidoi*. Signorelli and Scarabino (2010) tracked this misidentification to Ihering (1907).

Relevant literature

Paleontology: Figueiras, 1962: 58; Lopes et al., 2013; Parodiz, 1962: 42.

Taxonomy: Scarabino et al., 2006: 164; Signorelli and Pastorino, 2012a: 86; Signorelli and Scarabino, 2010: 34; Zelaya, 2016: 252. Cited as *Mactra patagonica* d'Orbigny by several authors: Barattini and Ureta, 1961: 169; Carcelles, 1944: 283; Castellanos, 1970: 235; Cauquoin, 1967: 225; Figueiras and Sicardi, 1969: 365; Ihering, 1907: 423; Rios, 1994: 265, 2009: 535.

Spisula (Spisula) isabelleana *(d'Orbigny, 1841) (Fig. 3.3a–d)*

Shell trigonal, with irregular growth lines on the external surface, covered almost entirely by a yellow to brownish and dehiscent periostracum; prosogyrous umbones present, not in contact; rudimentary lunule and escutcheon not defined by shelly

Fig. 3.3 (**a–d**) *Spisula (Spisula) isabelleana* d'Orbigny, 1846, NHMUK 1854.12.4.667/1, Lectotype; (**e–h**) *Spisula (Spisula) marplatensis* Doello Jurado, 1949, MACN-In 10307, holotype. Scale bar: (**a–h**) 2 cm

line; anterior and posterior dorsal margins softly curved and descended to rounded ends; ventral margin also curved; external ligament poorly developed, placed in a nymph oriented to the posterior side of the umbones; internal ligament trigonal, located in a ventrally developed chondrophore; interiorly white; hinge with two anterior lateral teeth (AI and AIII) in the right valve; two not fused cardinal teeth (3a and 3b) placed dorsally to the chondrophore and two posterior lateral teeth (PI and

PIII); left valve with one anterior (AII) and one posterior (PII) lateral tooth; inverted V-shaped cardinal tooth composed by two single cardinals (2a and 2b), the 2b flanked by accessory lamella (4b); left laterals with internal pustules well developed observed also in juveniles; external ligament poorly developed in juveniles; ultrastructure of the shell with two layers, the inner with a cross-lamellar complex and the outer with simple cross lamellar structure; juveniles with a protoconch very evident and a rudimentary hinge in both valves. Siphons type C of Yonge (1948, 1957), completely fused with outer ring of sensory tentacles, but without categorization; both siphons distinguished, the exhalant smaller than the inhalant; mantle folds united with the periostracal groove; ctenidia type C1 (Atkins 1937), inner demibranch with a food groove in the ventral margin; labial palps trigonal, elongated with the inner surface plicate and the outer smooth, fused to the ctenidia with a type 3 of Stasek category (1963); stomach type V of Purchon (1960), connected to a long crystalline style on the posterior ventral margin, completely separated from the intestine and located into a sac, immersed in the digestive diverticula; intestine long, coiled and positioned surrounding the visceral mass.

Type locality: Montevideo, Uruguay.

Type material: NHMUK 1854.12.4.667/1, lectotype, NHMUK 1854.12.4.667/2, 3, two paralectotypes, designated by Aguirre (1994).

Geographic distribution: Cabo Frio, Rio de Janeiro state, Brazil to San Antonio Oeste, Río Negro province, Argentina.

Stratigraphic distribution: Pleistocene—Recent.

Synonyms: None.

Remarks: *Spisula isabelleana* is one of the dominant bivalves in muddy and sandy bottoms in the Río de la Plata estuary (Giberto et al. 2004). Aguirre (1994) suggested the presence of a "polytypic" species that includes the morphotypes *S. petitii* and *S. marplatensis*. This position is not shared. The analysis of the type material revealed morphological differences to confirm the three species as valid. *S. isabelleana* differs from *S. marplatensis* by the presence of a more trigonal and less elongated shell; a shorter pallial sinus, stronger hinge with shorter lateral teeth. Moreover, *S. petitii* differs from *S. isabelleana* by the presence of a more fragile and less inflated shell.

Relevant literature

Anatomy and Morphology: Signorelli and Pastorino, 2006, 2012a.

Ecological and Environmental studies: Giberto et al., 2004; Signorelli et al., 2013.

Paleontology: Aguirre, 1989, 1991, 1992, 1994; Closs and Forti Esteves, 1971: 32; Parodiz, 1962: 36; Richards and Craig, 1963: 137.

Taxonomy: Adams and Adams, 1853–1858: 376; Barattini and Ureta 1961: 168; Carcelles, 1944: 283; Castellanos, 1970: 237; Cauquoin, 1967: 223; Figueiras and Sicardi, 1969: 365; Figueiras, 1961: 17; Lamy, 1917: 194; Rios, 1966: 33, 1970: 194, 1975: 235, 1985: 242, 1994: 265, 2009: 535; Scarabino et al., 2006: 157; Scarabino, 1977: 210; Scarabino, 2003: 245; Signorelli and Pastorino, 2012a: 88–90; Smith E. A., 1885: 58; Weinkauff, 1884: 115 (in Küster et al., 1841–1884); Zelaya, 2016: 253.

Spisula (Spisula) marplatensis *(Doello Jurado in Carcelles, 1949) (Fig. 3.3e–h)*

Shell oval, elongated, equilateral, strong, with the external surface smooth covered by a brownish periostracum usually absent in the umbonal area; umbones subcentral and prosogyrous; anterior end rounded ascending to the umbo straightly; ventral margin convex; posterior end with a more defined angle; escutcheon defined by a shelly ridge revealed by periostracal folds; lunule wide but poorly defined; interior white, with a deep pallial sinus, U-shaped; adductor muscle scars similar and lenticular; right hinge with two lateral teeth, different in length, being the AI more elongated than the AIII; chondrophore small and rounded; posterior lateral teeth more elongated than the anterior, similar in size and shape; cardinal teeth not fused, with the anterior short but strong, oriented obliquely to the anterior side of the shell and the posterior more fragile and placed almost vertically; left hinge with one anterior and one posterior lateral tooth similar in size and fragile; cardinal teeth fused making the typical inverted V-shaped tooth flanked by the accessory lamella; internal ligament in a resilifer, external ligament rudimentary located behind the umbones in the dorsal area; ultrastructure of the shell with same structure as *S. isabelleana*.

TYPE LOCALITY: Mar del Plata, Argentina in 30 m depth.

TYPE MATERIAL: MACN-In 10307, holotype, MACN-In 10307-1, six paratypes.

GEOGRAPHIC DISTRIBUTION: From Rio de Janeiro, Brazil to Puerto Quequén, Buenos Aires province, Argentina.

STRATIGRAPHIC DISTRIBUTION: Holocene—Recent.

SYNONYMS: None.

REMARKS: *Mactra marplatensis* was mentioned by Carcelles (1944) in his catalogue of marine mollusks from Puerto Quequén without a formal description. It was validated in Doello-Jurado in Carcelles (1949). The latter author added some notes to the unpublished manuscript of Doello-Jurado after the latter's decease in 1948. This species was commonly recorded along the northern coast of Buenos Aires province.

RELEVANT LITERATURE

Anatomy and Morphology: Signorelli and Pastorino, 2012a.

Paleontology: Parodiz, 1962: 42–43; Aguirre, 1994.

Taxonomy: Carcelles, 1944: 283 (*n. n.*); Castellanos, 1970: 234; Cauquoin, 1967: 224; Doello Jurado in Carcelles, 1949: 4; Figueiras and Sicardi, 1969: 366; Rios, 1966: 33, 1970: 195, 1975: 235, 1985: 243, 1994: 265; Signorelli and Pastorino, 2012a: 90–91; Zelaya, 2016: 253.

Spisula (Spisula) petitii *(d'Orbigny, 1841) (Fig. 3.4a, b)*

Shell trigonal to oval, inequilateral, elongate, maximum size to 45 mm, relatively fragile; external surface mostly smooth; concentric ornamentation only observed in the dorsal area at both sides of the umbones; posterior end rounded; posterodorsal

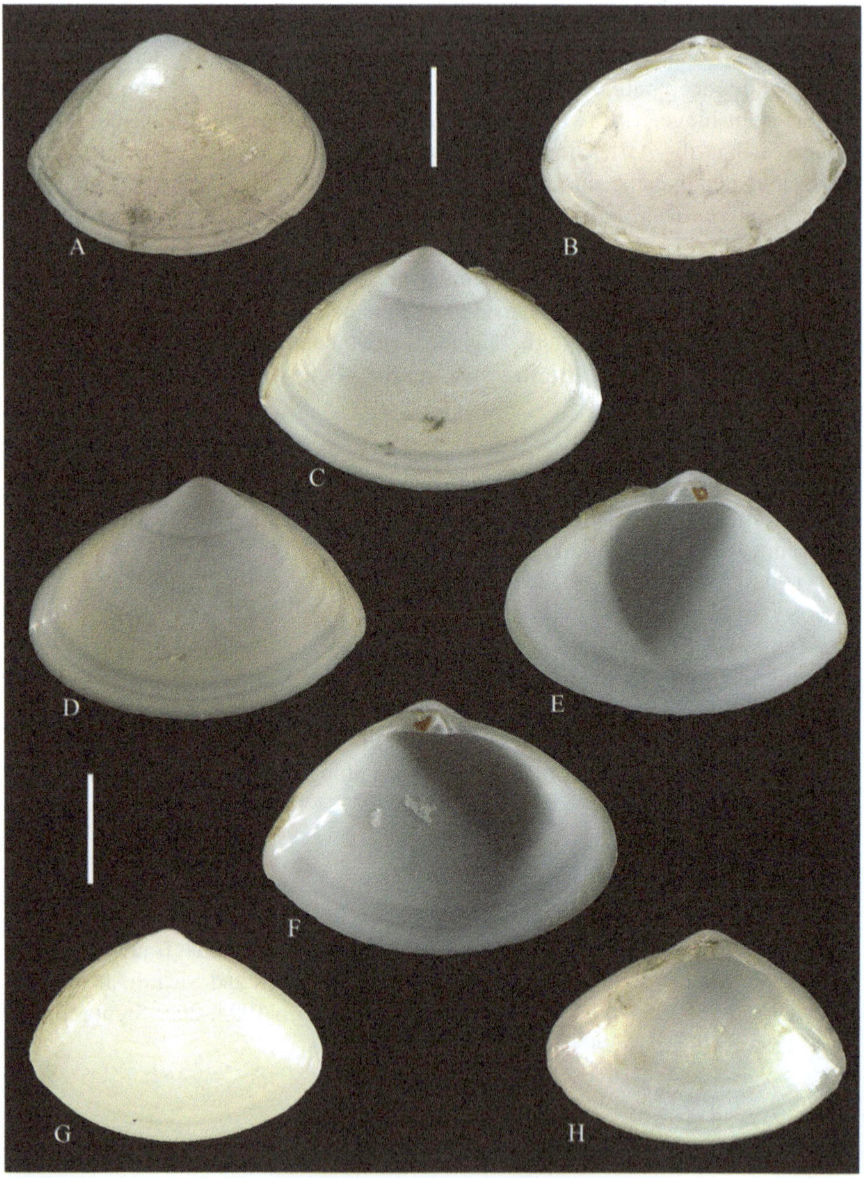

Fig. 3.4 (**a**, **b**) *Spisula (Spisula) petitii* d'Orbigny, 1846, NHMUK 1854-12-4-665, holotype; (**c**–**h**) *Spisula (Spisula) williamsi* (Berry, 1960), (**a**–**d**) SBNHM 213928, Jalisco, Mexico, (**e**, **f**) CAS 043985, Holotype. Scale bar (**a**–**h**) 1 cm

area or escutcheon well defined from the umbones to the posterior end, ventral margin and anterior end rounded without sharp angles; small and poorly inflated umbones at middle of the shell length.; interior white with a hinge plate similar to *S. williamsi*, with lateral teeth fragile and elongated, cardinal teeth symmetrically oriented in relation to the longitudinal axis, very close to the umbones; short but high U-shaped pallial sinus.

TYPE LOCALITY: Rio de Janeiro, Brazil.

TYPE MATERIAL: NHMUK 1854.12.4.665, holotype.

GEOGRAPHIC DISTRIBUTION: From Jamaica to southern Brazil.

STRATIGRAPHIC DISTRIBUTION: Recent.

SYNONYMS: *Mactra symmetrica* Reeve, 1854a: sp 84; and *Mactra richmondi* Dall, 1894c: 28.

REMARKS: *Spisula petitii* was described from Rio de Janeiro and is well known in southern localities (Cauquoin, 1967; Castellanos, 1970; Rios, 1970, 1975, 1985, 1994). No morphological differences were observed in the type material of *Mactra richmondi* (USNM 124774, syntype) from Nicaragua or *Mactra symmetrica* (NHMUK 20190674, holotype from Cuming collection). The southern distribution range of *Mactra petitii* was reviewed. The records for the Uruguayan and Argentinean coasts mentioned by Signorelli and Pastorino (2012a) belong to Holocene small specimens of *Spisula isabelleana* and *Spisula* sp. No records of living *petitii* were observed as suggested by Scarabino et al. (2016). *Mactra coquimba* was mentioned by Cauquoin (1967: 226) as a junior synonym of *S. petitii*. However, Cauquoin misspelled the species. The correct name is *Mactra coquimbana* Philippi, 1887, from Coquimbo. After the examination of type material (MNHNS 212, holotype), this nominal species is, in fact, a junior synonym of *Mulinia edulis* (Signorelli and Pastorino, 2011). Finally, *Mactra scalpellum* Reeve, 1854a (NHMUK 20190675, syntype) from New Zealand coast was erroneously mentioned by Dall (1891) as living in Brazil. The type material of this species clearly differs from *S. petitii*.

RELEVANT LITERATURE

Paleontology: Aguirre, 1994; Ihering, 1907: 320.

Taxonomy: Altena 1971: 52; Castellanos, 1970: 234; Cauquoin, 1967: 226; Conrad, 1868: 39; Dall, 1891: 44; 1894c: 26; Lamy, 1918: 342; Martinez and Saez, 1869: 13; Palacio et al., 2010: 406; Rios, 1970: 196, 1975: 236, 1985: 243, 1994: 265, 2009, 536; Signorelli and Pastorino, 2012a, 91–92; Zelaya, 2016: 253.

Spisula (Spisula) williamsi *(Berry, 1960) (Fig. 3.4c–h)*

Shell ovate-trigonal, medium size, length to 36 mm, thick; umbo small, moderately inflated, pointed; external surface smooth with commarginal undulations; anterior end slightly elongated, rounded but sharper than posterior end; commarginal ribs along the anterodorsal and posterodorsal areas; periostracum thin, yellowish, shiny; posterodorsal margin slightly rounded behind the umbo; external ligament short; internally white, porcelaneous; hinge plate with fragile but elongated lateral teeth, cardinal teeth large, chondrophore moderately projected ventrally; pallial sinus moderate to deep, about 40% of shell length, U-shaped.

Type locality: off La Libertad, Guayas, Ecuador, 18 m depth.

Type material: CAS 043985, holotype dry; SBMNH 34019, paratype 1 valve, dry (previous number SSB 9849), collected in 1937.

Geographic distribution: From Bahía Santiago, Colima, Mexico to La Libertad, Guayas, Ecuador. Reported from Zorritos, Tumbes, Peru by Olsson (1961, as *Spisula adamsi*) and from Atacames, Ecuador by Cauquoin (1969a, as *Mactra hoffstetteri*).

Stratigraphic distribution: Holocene—Recent.

Synonyms: *Spisula adamsi* Olsson, 1961: 326; *Mactra (Simomactra) hoffstetteri* Cauquoin, 1969a: 1021.

Remarks: Type material examination of *Spisula adamsi* (ANSP 218917, holotype) and *Mactra (Simomactra) hoffstetteri* (MNHN-IM-2000-34899, syntypes) confirmed both species as junior synonyms of *S. williamsi* as suggested by Coan and Valentich-Scott (2012). Shell outline and hinge plate morphology suggest a subgeneric affinity with *S. petitii* that must be genetically tested.

Relevant literature

Taxonomy: Cauquoin, 1969a: 1021–1023; Coan and Valentich-Scott, 2012: 555–556; Keen, 1971: 202; Olsson, 1961: 326; Paredes et al., 2016: 141.

Subgenus **Hemimactra** *Swainson, 1840: 369*

Shell large, oval to trigonal, smooth, with dorsal areas not defined; anterior cardinal teeth in RV aligned with the lateral teeth; laterals with pustules on the dorsal surface.

Type species: *Mactra gigantea* Lamarck, 1818: 472 (= *M. solidissima* Dillwyn, 1817: 140) by monotypy.

Geographic distribution: East North America.

Remarks: Several works recognized two valid species included in the subgenus *Hemimactra* as living in the east coast of the United States (Porter and Schwartz, 1981; Vecchione and Griffiths, 1996; Mikkelsen and Bieler, 2007; Turgeon et al., 2009). Recently, Hare and Weisberg (2005) based on mitochondrial and nuclear DNA sequences, confirmed the presence of two valid species, one from the northern coast of United States (*S. (Hemimactra) solidissima*) and the second from Cape Code bay to the Gulf of Mexico (*S. (Hemimactra) raveneli*).

Spisula (Hemimactra) solidissima (Dillwyn, 1817) (Fig. 3.5a–d)

Shell ovate, sub-triangular, solid, thick, elongated, large, with a reported shell length of 226 mm; umbo slightly anteriorly placed, prosogyrate; moderately inflated; external surface smooth, with small irregular growth lines; with a thin yellowish brown periostracum; anterodorsal margin gently curved; anterior end rounded,

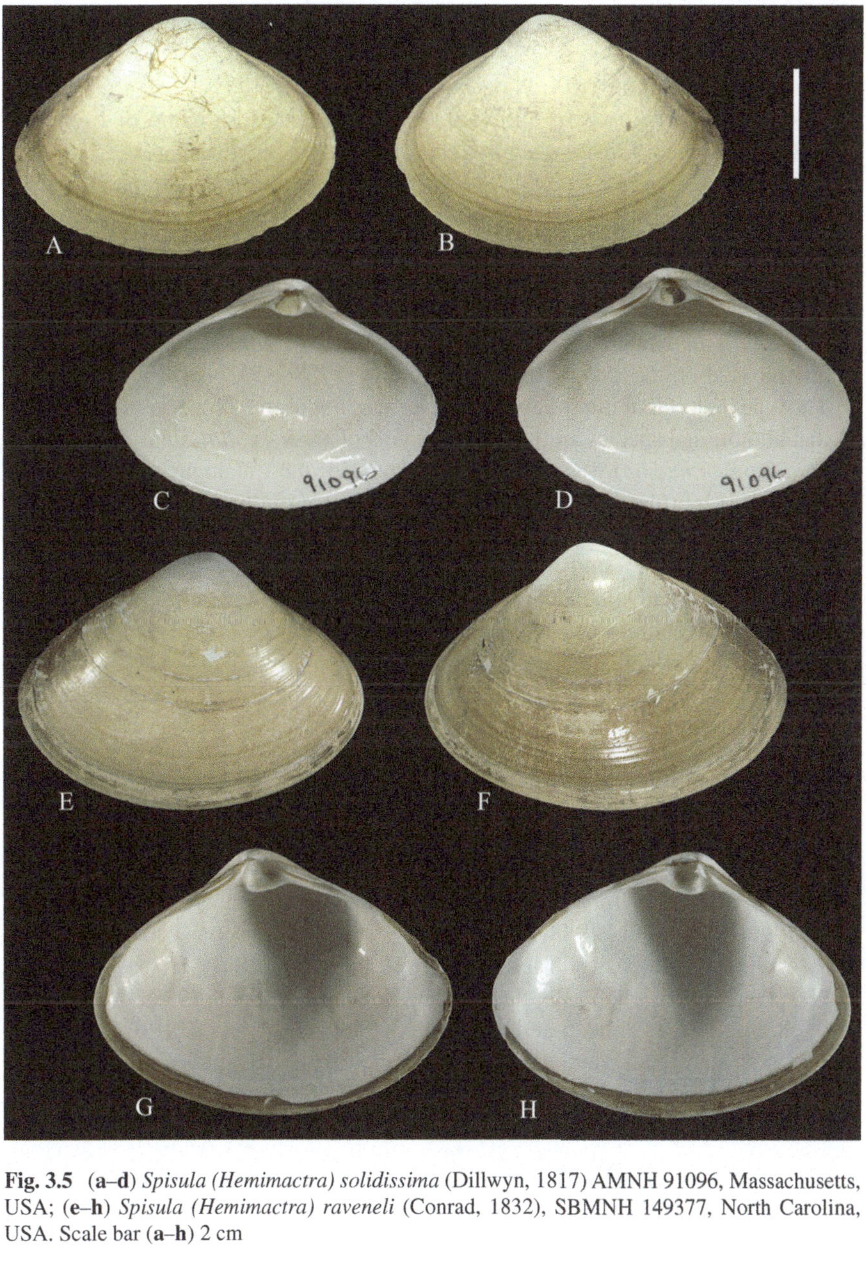

Fig. 3.5 (**a–d**) *Spisula (Hemimactra) solidissima* (Dillwyn, 1817) AMNH 91096, Massachusetts, USA; (**e–h**) *Spisula (Hemimactra) raveneli* (Conrad, 1832), SBMNH 149377, North Carolina, USA. Scale bar (**a–h**) 2 cm

posterodorsal margin straight, long; posterior end rounded, more acute that anterior; internally glossy, white, hinge plate nearly central, placed slightly near the posterior end, with striated lateral teeth; chondrophore trigonal, ventrally projected, large; pallial sinus short, U-shaped.

Type locality: North America was mentioned by Chemnitz (1788).

Type material: Not found, name based on Chemnitz specimens (1788: 350, pl. 170, figure 1656, work rejected by ICZN, 1944, Opinion 184).

Geographic distribution: From Nova Scotia, Canada to North Carolina, United States.

Stratigraphic distribution: Pliocene—Recent.

Synonyms: ?*Mactra procera* Dillwyn, 1817: 140 (based on name of Lightfoot, 1786: 24, Solander ms, *nomen nudum*); *Mactra gigantea* Lamarck, 1818: 472; *Mactra similis* Say, 1822: 309; ?*Spisula solidissima acadica* Matthew, 1874: 42; ?*Spissula solidissima peninsulae* M. Smith, 1937: 65–66.

Remarks: This species constitutes one of the most studied mactrid in American waters. Several aspects have been studied, such as fishery, taxonomical or ecological (see Relevant Literature section). In relation to the junior synonyms of *S. (H.) solidissima* mentioned in the literature, *Mactra gigantea* is an objective synonym based on the same figure of Chemnitz (1788: 350, pl. 170, figure 1656). The type material of the following species was not found, however, it is probable that all nominal species are junior synonyms of *S. (H.) solidissima. Mactra procera* (*nomen nudum*) was based on specimens of the Catalogue of the Portland Museum (Lightfoot, 1786) from New York. *Mactra similis* (types not found) was collected from the coast of New Jersey. In addition to that, *Spisula solidissima acadica* (types not found) was described from Labrador au détroit, Long Island (New York). Finally, the identity of *Spissula solidissima peninsulae* (types in the personal collection of the author) is uncertain. It was described as a fossil from Pliocene deposits of Clewiston, Florida.

Relevant literature (*Spisula solidissima* is one of the most studied mactrid, a selected literature is provided in this section)

Anatomy and Morphology: Cable, 1973; Dean, 1980; Ivany et al., 2003; Jacobson, 1972a, 1972b; Jones et al., 1978; Longo and Anderson, 1969; Patton and Quinn, 1973; Prior, 1974; Rebhun, 1962; Ropes and O'Brien, 1980; Russell-Hunter and Grant, 1962.

Ecological and Environmental studies: Cargnelli et al., 1999; Cerrato and Keith, 1992; Jones et al., 1983; Jones, 1980, 1981; Narváez et al., 2015; Robinson et al., 1984; Ropes and Merrill, 1966, 1973; Savage, N. B. 1976.

Fisheries: Allen, 1953; Belding, 1910; Bell and Grassle, 1998; Cessine and Strand, 1978; Clotteau and Dubé 1993; Engle and Merrill, 1967; Goldberg, 1989; Hare and Weinberg, 2005; Hare et al., 2010; Loosanoff and Davis, 1963; Munroe et al., 2013; Ropes, 1968a, 1968b; Ropes et al., 1969; Walker and O'Beirn, 1996.

Physiology and Genetics: Krantic et al., 1991; Maul and Avdalović, 1980; Palazzo et al., 1992; Rosenthal et al., 1980; Schuetz, 1975; Stecher et al., 1996; Vogel et al. 1997; Weisenberg, 1972.

Taxonomy: Abbott, 1974, 1991; Brunel et al., 1998: 133; Dall, 1894c: 28; Gould, 1841: 51; Huber, 2010: 448. Jacobson and Old, 1966; Lindell et al., 2006; Matthew, 1874: 42; Mikkelsen and Bieler, 2007: 372; Morris, 1951: 97; Yancey and Welch, 1968; Zim and Ingle, 1955: 156.

Spisula (Hemimactra) raveneli (Conrad, 1832) (Fig. 3.5e–h)

Shell rounded, trigonal, solid, large, length to 125 mm; external surface smooth with a concentric and irregular growth lines; anterodorsal margin slightly convex, anterior and posterior ends rounded, posterodorsal area straight defined by wrinkles of the periostracum which is yellowish brown, more conspicuous over the ventral edge; internally white; hinge plate with long lateral teeth, V-shaped cardinal teeth formed by the fusion of 3a and 3b cardinal tooth, the 3a anteriorly elongated; chondrophore well ventrally developed; pallial sinus moderate, V-shaped, about one-third shell length.

Type locality: Charleston, South Carolina, United States, designated by Jacobson and Old (1966).

Type material: Not found, not designated in the original description.

Geographic distribution: East coast of North America from Massachusetts, United States to Gulf of Mexico, Mexico.

Stratigraphic distribution: Pleistocene—Recent.

Synonyms: *Spisula sayii* Gray, 1837: 373.

Remarks: *Spisula (Hemimactra) raveneli* (Conrad, 1832) was described as living in the coast of Virginia, United States, distinguished from *S. solidissima* by having a smaller size, a southerly distribution, and a more elongated shell. The taxonomic position of this species was widely discussed by several authors who considered it as a junior synonym of *Spisula similis* Say, 1822 (Perry, 1940; Abbott 1974; Emerson and Jacobson 1976; Cargnelli et al., 1999, among others). However, Jacobson and Old (1966) suggested that Say described a pathologic form of *Spisula solidissima* from New Jersey. And the southern populations referred to as *S. similis* belong, in fact, to *S. raveneli*.

Relevant literature

Ecological and Environmental studies: Cargnelli et al., 1999; Hare and Weinberg, 2005; Hare et al., 2010; Porter and Schwartz, 1981; Williams and Porter, 1971.

Fisheries: Goldberg, 1989; Walker and Heffernan 1994.

Taxonomy: Abbott, 1974; Andrews 1971; Camp et al., 1998: 12; Cooley, 1978; Dall, 1889, 1894c: 26; Huber, 2010: 449; Jacobson and Old, 1966: 31; Jacot, 1920: 137; Lamy, 1918: 316–317; Lermond, 1936; Merril and Ropes, 1969; Mikkelsen and Bieler, 2007: 372; Prezant et al., 2002: 349; Turgeon et al., 2009: 730; Vecchione and Griffis, 1996.

Genus *Standella* Gray, 1853

[= *Mactrotoma* Dall, 1894c, type: *Mactra fragilis* Gmelin, 1791, by original designation; *Papyrina* Mörch, 1853 in part]

Shell elongate, posteriorly truncated and gaped, posterodorsal area defined by a shelly line; umbo smooth or with commarginal undulations; lateral teeth not pustulose.

TYPE SPECIES: *Mactra fragilis* of Chemnitz, 1782, rejected work (= *Mactra fragilis* Gmelin, 1791), by original designation.

GEOGRAPHIC DISTRIBUTION: Pacific and Atlantic coast of America, South Africa, Indo Pacific from West Africa to Japan.

REMARKS: Historically, the genus *Mactrotoma* was used instead of *Standella* due to the misinterpretation of the type species. The taxonomic status of both genera was recently revised (Petit, 2012; Huber, 2015) where two points of view were stated. On the one hand, Petit (2012) suggested that both genera are based on the same type species and therefore must be considered synonyms. But on the other hand, Dall (1898a), mentioned that the type species of *Standella* is *Spisula fragilis* Gray, 1853 (= *Mactra pellucida* Gmelin, 1791, based on the figure 234 of Chemnitz, 1782) by subsequent designation. Huber (2015) followed Dall (1898a) and considered that type species of *Mactrotoma* is *Mactra fragilis* Gmelin, 1791, name based on the figure 235 of Chemnitz, 1782. After the review of all related literature (Dall, 1898a, Lamy, 1917–1918, Huber, 2010, 2015, Petit, 2012), Petit's argument is shared. When the genus *Standella* was described by Gray (1853), the author mentioned the species *S. fragilis* and *S. aegyptiaca*, without authorities or dates, and without designation of type species. Although type designation is not affected, no evidence supports Dall' conclusion. Within this context, the type species of both genera is *Mactra fragilis* Gmelin, 1791, name based on the figure 235 of Chemnitz (1782). This conclusion was also confirmed by Valentich-Scott, Coan and Zelaya (in prep.). Therefore, eight valid species belonging to the genus *Standella* are recognized as living in American waters. Besides the type species of the genus, the other truly *Standella* is *S. (Standella) nasuta* (Gould, 1851). The taxonomic status of the genera *Micromactra* Dall, 1894d and *Mactroderma* Dall, 1894d is unresolved, and they are herein considered as subgenera of *Standella*. Undoubtedly, genetic studies are needed. The synonymy between *Micromactra* and *Mactrella*, proposed by Signorelli (2012), was rejected by Huber (2015) who considered *Micromactra* as a subgenus of *Mactrotoma*. After the re-examination of types and biogeographical data, the opinion of Huber is shared. In the subgenus *Micromactra*, four species are included as living in American waters. They are the type species *Standella (Micromactra) californica*, *S. (M.) angusta* (Reeve, 1854a), *S. (M.) janeiroensis* (E. A. Smith, 1915) and *S. (M.) surinamensis* (Altena, 1968). The last two species, very similar in shell shape, were considered synonyms by Signorelli and Pastorino (2012a) but separated by Huber (2015) due to a geographical gap and shell differences. Although these morphological differences would be considered as intraspecific, we keep both as valid species awaiting molecular confirmation. Finally, in the subgenus *Mactroderma* two valid species are recognized: *S. (Mactroderma) isthmica* (Pilsbry and Lowe, 1932) and *S. (M.) velata* (Philippi, 1849).

Subgenus Standella

Standella (Standella) fragilis (Gmelin, 1791) (Fig. 3.6a–d)

Shell trigonal to oval, inequilateral, external surface smooth without ornamentation, posterodorsal area defined by a line from umbones to posterior end and covered by a brownish periostracum; anterior end rounded, lunule not defined; ventral margin convex; internally white; right hinge with two anterior (AI and AIII) and two posterior (PI and PIII) lateral teeth, the ventral anterior with two cusps, the posterior similar in size and with one cusp; two cardinal teeth (3a and 3b) unfused; left hinge with V-shaped cardinal tooth composed of two single cardinals (2a and 2b), one anterior and one posterior lateral tooth (AII and PII), the anterior with two cusps; ultrastructure of the shell with two layers, an outer crossed-lamellar and an inner complex crossed-lamellar structure.

Type locality: Nicobar Islands, India, erroneously registered.

Type material: Types not found at ZMUC-BIV, ZMB and Zoological Institute of Russian Academy of Science of St. Petersburg (Martynov, 2002), probably lost.

Geographic distribution: From North Carolina, United States to Santa Catarina, Brazil.

Stratigraphic distribution: Lower Pleistocene—Recent.

Synonyms: *Lutraria candida* Lamarck, 1818: 470; *Mactra braziliana* Lamarck, 1818: 478; *Mactra oblonga* Say, 1822: 310; *Mactra bilineata* Reeve, 1854a: sp. 72; *Mactra anserina* Guppy, 1875: 49; ?Mactra *fragilis var. gigas* Nowell-Usticke, 1969: 7.

Remarks: *Mactra fragilis* was described by Chemnitz (1782: 236, pl. 24, fig. 235). However, this work was rejected due to nomenclatorial purposes (ICZN 1944, Opinion 184). It was attributed to Gmelin (1791). This is a common living species of the western Atlantic. Type material examination of *Lutraria candida* (MNHN-IM-2000-7036, two syntypes); *Mactra braziliana* (MNHN-IM-2000-5015, one syntype); *Mactra oblonga* (ANSP 51418, one syntype) and *Mactra bilineata* (NHMUK 20190676/1, /2, two syntypes) confirmed all nominal species as junior synonyms of *S. fragilis*. The types of *Mactra anserina* were not found at NHMUK or Guppy main collection deposited at Victoria Institute, Trinidad, destroyed by fire in 1920 (Sherborn 1940). However, this species is herein considered as a synonym according to the similarities observed in the original illustration. *Mactra oblonga* was erroneously cited by Ravenel (1834) as *M. oblongata* (*nomen nudum*). *Mactra dealbata* Pulteney, 1799, later named as *M. tellinoides* Pulteney, 1813, together were considered synonyms of

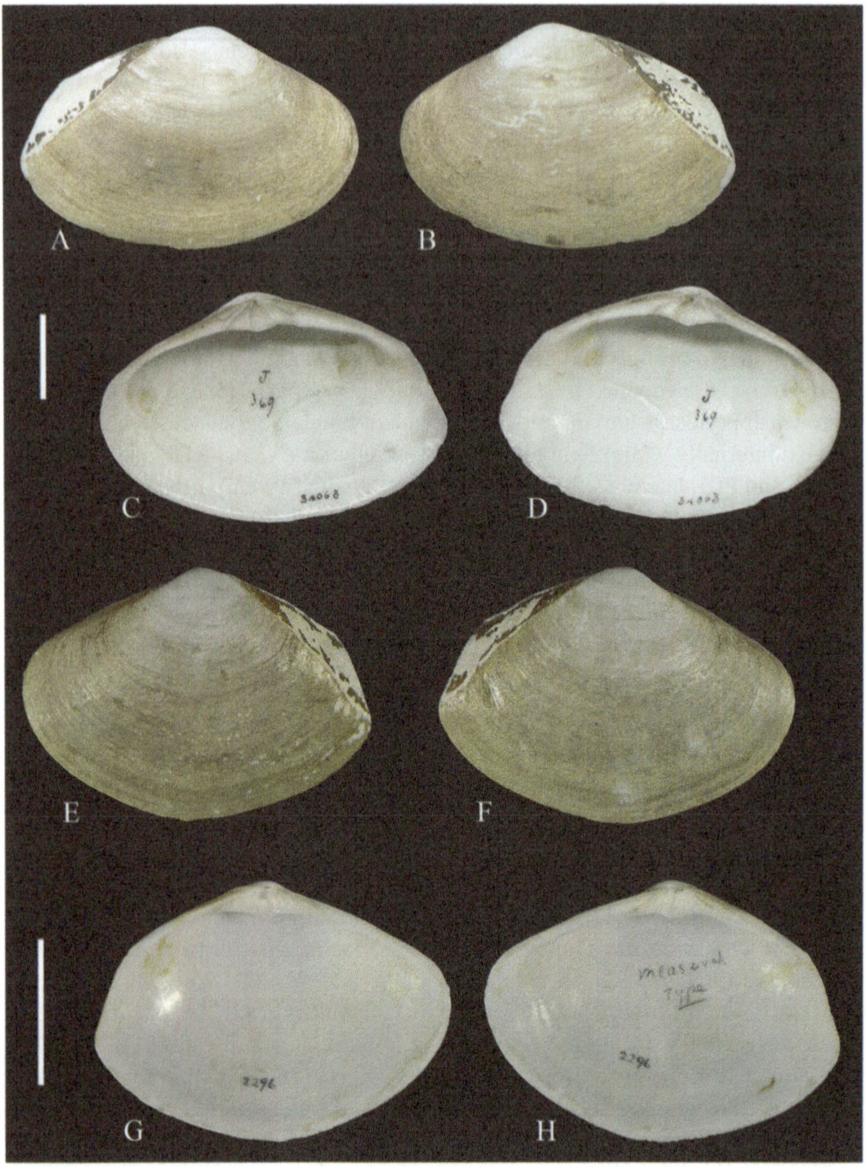

Fig. 3.6 (**a–d**) *Standella (Standella) fragilis* (Gmelin, 1791), AMNH 34063; (**e–h**) *Standella (Standella) nasuta* (Gould, 1851), USNM 2296, syntypes. Scale bar (**a–d**) 2 cm, (**e–h**) 3 cm

S. fragilis by several authors (Wood and Hanley, 1856; Dall, 1894c; Lamy, 1917–1918; Huber, 2010, among others). The identity of these names was revealed by Oliver et al. (2017: 379) who found potential types (NHMUK 20170127, 2 valves). *Mactra dealbata* and *M. tellinoides* must be regarded as junior synonyms of *Standella pellucida* (Gmelin, 1791) from Indo-Pacific. *Mactra ambigua* Weinkauff, 1884 (in Küster et al., 1841–1884), mentioned by Dall (1894c) as a synonym of *S. fragilis*, is in fact a synonym of *S. compressa* Spengler, 1802, living along the northern Atlantic coast of Africa. And *Mactra ovalina* Lamarck, 1818, cited as a synonym of *S. fragilis* by Carpenter (1855, in 1855–1857), Dall (1894c), and others, is a valid species living in Arabia (Oliver, 1992; Bosch et al. 1995). Finally, *Mactra fragilis var. gigas* was not illustrated in the original description. The figured specimen given by Nowell-Usticke (1971, pl. 5, fig. 1784A) suggested affinities with *Standella fragilis*. Faber (1988) only found a label in AMNH cabinets and considered the types as lost. *Mactra fragilis var. gigas* must be regarded as a *nomen dubium*.

RELEVANT LITERATURE

Ecological and Environmental studies: Abraham et al., 2008.
Paleontology: Gardner and Mansfield, 1943.
Taxonomy: Adams and Adams, 1856 in 1853–1858: 382; Altena, 1971: 52; Camp et al., 1998: 12; Carpenter, 1855, in 1855–1857: 51; Cauquoin, 1967: 223; Conrad, 1832: 61, 1868: 45; d'Orbigny, 1846 (in 1834–1847): 508; Dall, 1894c: 26, 1895b: 211; De Kay, 1843: 230; Deshayes and Milne-Edwards, 1835: 92; Gmelin, 1791: 3261; Gondim et al., 2011: 77; Gray, 1837: 373; Gundlach, 1883b: 482; Guppy, 1875: 49; Huber, 2010, 2015; Lamarck, 1818 in 1815–1822: 470, 478; Lamy, 1917: 246; Martinez and Saez, 1869: 12; Maury, 1920: 131; Perry, 1940: 82; Reeve, 1854a: sp 47; Reguero and Raz-Guzmán, 2018: 36; Rios, 1970: 195, 1975: 236, 1985:243, 1994: 265, 2009: 535; Robinson and Montoya, 1987: 396; Say, 1822: 310; Signorelli and Pastorino, 2012b: 38–42; Spengler, 1802: 124; Tallarico et al., 2014: 101; Tomlin, 1924: 136; Turgeon et al., 2009: 730; Weinkauff, 1882: 84 (in Küster et al., 1841–1884); Wood and Hanley, 1856: 39; Wood, 1818: 30.

Standella (Standella) nasuta (Gould, 1851) (Fig. 3.6e–h)

Shell trigonal to rhomboidal, equilateral, medium size to large, length to 120 mm; slightly anteriorly elongated, thin, moderately inflated; umbo small, not inflated; external surface smooth, with commarginal striae; anterodorsal margin straight; anterior end rounded, ventral margin convex; posterodorsal slope wide, set off by shelly ridge; moderately gaped; brownish and thin periostracum, dark brown on posterior slope; internally white; hinge plate wide, long; with a broad space between chondrophore and posterior lateral teeth; chondrophore slightly ventrally projected; pallial sinus broad and deep, U-shaped, about 45% of shell length.

TYPE LOCALITY: Mazatlán, Mexico.
TYPE MATERIAL: USNM 2296, Syntypes, two articulated specimens.
GEOGRAPHIC DISTRIBUTION: from Santa Barbara, California, United States to Tumbes, Peru.

STRATIGRAPHIC DISTRIBUTION: Upper Pleistocene—Recent.

SYNONYMS: *Mactra californica* Reeve, 1854a: pl. 20, sp 114, ex Deshayes MS (*non* Conrad, 1837); *Mactra hiantina* Deshayes, 1855a: 68; *Mactra deshayesii* Conrad, 1868: 46 (*non* Mayer, 1867); *Mactrotoma revellei* Durham, 1950: 93.

REMARKS: Similar to the west Atlantic *S. (S.) fragilis*, but with the anterior end longer, more inflated and rounded ventral margin. The illustration of *Mactra californica* given by Reeve (1854a: pl. 20, sp 114) fits more adequately to *S. (S.) nasuta* than with *S. (Micromactra) californica* of Conrad, 1837. For this reason, it is probable that Conrad (1868: 46) renamed Reeve's californica with *Mactra deshayesii*. A complete specimen is deposited into the cabinets of NHMUK, and it is probably the type of Conrad's new name. *Mactra hiantina*, described from Puna, Guayaquil, was usually mentioned as a junior synonym of *S. (S.) nasuta* (Huber, 2010; Coan and Valentich-Scott, 2012). However, the types were not found to confirm this status. Finally, *Mactrotoma revellei* is confirmed as a synonym after the examination of type material (UCMP 32237, holotype).

RELEVANT LITERATURE

Paleontology: Packard, 1916: 278–279.

Taxonomy: Coan and Valentich-Scott, 1997: 17, 2012: 550; Dall, 1916a: 39; Everson, 1991; Hertlein and Strong, 1950: 229–231; Huber, 2010: 444; Keen, 1971: 202; Paredes et al., 2016: 141; Poorman and Poorman, 1978: 371; Strong, 1925.

Subgenus Micromactra *Dall, 1894d*

Shell ovate-elongate similar to *Standella* but with concentric undulations near the umbo, moderately fragile, pallial sinus deep, broad.

TYPE SPECIES: *Mactra californica* Conrad, 1837, by monotypy.

GEOGRAPHIC DISTRIBUTION: West North to Central America, East Central to South America.

REMARKS: The taxonomic placement of the subgenus *Micromactra* was herein revised. After a reexamination of type species and additional material, the synonymy with *Mactrella* proposed by Signorelli (2012) is corrected. Although, the morphological characters of *Standella* and its included subgenera are similar, this relationship must be tested genetically. At the moment, *Micromactra* is considered as a subgenus of *Standella*. In this subgenus, four living species in American waters are recognized. They are *Standella (Micromactra) angusta* (Reeve, 1854a), *S. (M.) californica* (Conrad, 1837); *S. (M.) janeiroensis* (E. A. Smith, 1915) and *S. (M.) surinamensis* (Altena, 1968). The latter two are widely mentioned as synonyms (Rios, 1994, 2009; Signorelli and Pastorino, 2012b and literature therein).

Standella (Micromactra) angusta (Reeve, 1854a) (Fig. 3.7a–c)

Shell ovate, elongate, fragile, thin, equilateral, length to 55 mm; anterodorsal margin slightly concave; umbo narrow, not inflated, prosogyrate, posterodorsal margin straight; posterior and anterior ends rounded, the anterior one broader; external surface with commarginal undulations more evident near the umbo; escutcheon narrow, posterior slope defined by a low ridge; moderate posteriorly gaped, light brown periostracum; internally white, short and narrow hinge plate with the anterior lateral

Fig. 3.7 (**a–c**) *Standella (Micromactra) angusta* (Reeve, 1854a), NHMUK 20190677, holotype; (**d–g**) *Standella (Micromactra) californica* (Conrad, 1837), AMNH 34670, from San Pedro bay. Scale bar: 2 cm

teeth very close to the cardinals, posterior lateral teeth more elongated; chondrophore slightly ventrally projected; pallial sinus moderate, about 40% shell length.

TYPE LOCALITY: Panama.
TYPE MATERIAL: NHMUK 20190677, holotype.
GEOGRAPHIC DISTRIBUTION: West coast of America from Sinaloa, Mexico to Tumbes, Peru.
STRATIGRAPHIC DISTRIBUTION: Pliocene—Recent.
SYNONYMS: *Mactra (Micromactra) atacama* Pilsbry and Olsson, 1941: 73.
REMARKS: type material examination of *Mactra (Micromactra) atacama* (ANSP 175548, paratype) confirms it as a junior synonym of *S. (M.) angusta*. This species differs from *S. (M.) californica* by the presence of a concave anterodorsal margin and a longer anterior cardinal tooth in the left valve.
RELEVANT LITERATURE

Paleontology: Pilsbry and Olsson, 1941: 73.
Taxonomy: Cauquoin, 1969a: 1024; Coan and Scott, 2012: 549; Hertlein and Strong, 1950: 231, 1955: 203–204; Huber, 2010: 444; Keen, 1971: 204; Olsson, 1961: 325; Paredes et al., 2016: 141; Pilsbry and Lowe, 1932: 89; Ramírez et al., 2003: 268.

Standella (Micromactra) californica (Conrad, 1837) (Fig. 3.7e–g)

Shell sub-trigonal, elongate, equilateral, medium size, length to 55 mm; anterodorsal margin straight; umbo not inflated, narrow, posterodorsal margin straight; anterior end rounded, posterior end moderately truncate, narrowly gaped; external surface with fine commarginal striae, turning to commarginal undulations on the umbonal area; escutcheon narrow, posterior slope defined by a low ridge, brownish periostracum; internally white, hinge plate longer than in *M. angusta*, with the chondrophore not ventrally projected and the lateral teeth larger; external portion of ligament moderately developed; pallial sinus deep and broad, about 50% shell length.

TYPE LOCALITY: Santa Barbara, California, United States.
TYPE MATERIAL: Not found, probably lost (Keen, 1966).
GEOGRAPHIC DISTRIBUTION: West coast of America from California, United States, thorough Gulf of California, Mexico to Guayas, Ecuador.
STRATIGRAPHIC DISTRIBUTION: Upper Pleistocene—Recent.
SYNONYMS: None.
REMARKS: This is the type species of the subgenus and a well-known species of the west coast of North America. For this reason, given that its type material is probably lost, a designation of a neotype is not necessary according to the article 75.2 and 75.3 of the code. No synonyms were found in the literature.
RELEVANT LITERATURE

Paleontology: Grant and Gale, 1931: 392–393; Packard, 1916: 280–281.

Taxonomy: Carpenter in Lea et al., 1860: 2; Cauquoin, 1969a: 1024–1025; Coan and Scott, 1997: 17, 2012: 549; Coan and Valentich-Scott in Carlton, 2007: 848; Coan et al., 2000: 457; Conrad, 1868: 44; Dall, 1894d: 40, 1916a: 39; 1925: 19; Haderlie and Abbott, 1980: 378; Huber, 2010: 444; Keen, 1971: 204; López de Mesa and Cantera, 2015: 8; Poutiers, 1995: 144; Signorelli, 2012; Strong, 1925.

Standella (Micromactra) janeiroensis (E. A. Smith, 1915) (Fig. 3.8a–d)

Shell sub-trigonal, elongated, thin, fragile, inequilateral, with concentric ornamentation weaker along the ventral margin; subequilateral, umbones prosogyrous and weakly inflated; anterior end posterior rounded, posterior end pronounced; brownish periostracum eroded in umbonal area; lunule not defined by a shelly line but recognized; escutcheon well developed and limited by a line or a soft keel; interior white and pellucid; right hinge with two cardinal teeth (3a and 3b) not fused; anterior lateral teeth of different sizes, the ventral one longer; two posterior lateral teeth (PI and PIII) similar in size; left valve with inverted V-shaped cardinal teeth (2a and 2b) flanked in the posterior side by the accessory lamella (4b); one anterior and one posterior lateral tooth observed; chondrophore well developed ventrally; pallial sinus deep, U-shaped; ultrastructure of the shell with two layers, an outer crossed-lamellar and an inner complex crossed-lamellar structure; siphons completely fused covered by the periostracum, inhalant with three categories of tentacles; exhalant siphon with two rows of tentacles of uniform size.

Type locality: Rio de Janeiro, Brazil (station 42 of "Terra Nova" Antarctic Expedition).

Type material: NHMUK 1915.4.18.489, 6 Syntypes.

Geographic distribution: From Maceió, Brazil to Villarino beach, San José gulf, Argentina.

Stratigraphic distribution: Pleistocene—Recent.

Synonyms: None.

Remarks: The distribution of *Standella (Micromactra) janeiroensis* given by Signorelli and Pastorino (2012a) is restricted from Brazilian coast to Argentina. Huber (2015) suggested that the Caribbean *S. (M.) surinamensis* must be considered valid due to morphological differences of the umbonal undulations and the geographical gap between Caribbean and Brazilian-Argentinean populations. *Micromactra miskito* Petuch, 1998 described from the coast of Nicaragua, mentioned by Signorelli and Pastorino (2012a) as a synonym of *janeiroensis*, must be attributed to *surinamensis*. However, the morphological differences mentioned by Huber (2015) could be intraspecific. And the presence of only one species with an anti-tropical distribution must be genetically confirmed. At the moment, two valid species are accepted.

Relevant literature

Anatomy and Morphology: Signorelli and Pastorino, 2012a.

Fig. 3.8 (**a–d**) *Standella (Micromactra) janeiroensis* (E. A. Smith, 1915), NHMUK 1915.4.18.489, syntypes; (**e–h**) *Standella (Micromactra) surinamensis* (Altena, 1968), (**e**, **f**) Holotype, from Surinam, no exact locality, (**g**, **h**) Paratype from Popogaimama Creek. Scale bar: (**a–d**) 2 cm, (**e–h**) 5 cm

Paleontology: Figueiras, 1962: 62.
Taxonomy: Barattini and Ureta, 1961: 169; Castellanos, 1970: 236; Cauquoin, 1967: 224; Doello-Jurado in Carcelles, 1944: 284; Figueiras and Sicardi, 1969: 367; Rios, 1969: 9, 1970: 195, 1975: 236, 1985: 243, 1994: 266, 2009: 536; Signorelli and Pastorino, 2012a: 93; Signorelli, 2012; Zelaya, 2016: 253.

Standella (Micromactra) surinamensis (Altena, 1968) (Fig. 3.8e–h)

Shell sub-trigonal, elongated, solid, inequilateral, with concentric smooth undulation, more evident on the umbonal area; subequilateral; umbones prosogyrous; anterior and posterior ends rounded; lunule not observed, periostracum completely eroded in the umbonal area; escutcheon limited by a line well defined; interior white; right hinge with two cardinal teeth (3a and 3b) not fused; anterior lateral teeth (AI and AIII) of different sizes; two posterior lateral teeth (PI and PIII) similar in size; left hinge with inverted V-shaped cardinal teeth (2a and 2b) flanked in the posterior side by the accessory lamella (4b); one anterior and one posterior lateral tooth observed; chondrophore moderately ventrally projected; pallial sinus deep, U-shaped.

Type locality: Surinam, no exact locality given.
Type material: RMNH.MOL.55788, holotype, a left valve (Fig. 151a-c) from Surinam (no exact locality, but the specimen is obviously Recent), June 1910, collected by D. G. J. Bolten. Seven paratypes mentioned in the original description. Holocene paratypes (three lots): RMNH.MOL.55794, one right valve from Holocene deposits from Coronieweg, picket, Coronie District, collected by D. C. Geijskes; RMNH.MOL.55793, one left valve (figured 151d) from Coronieweg, km 117–118, Coronie District, collected by C. O v. R. Altena; RMNH.MOL.55798, one small right valve from Charlesburg, Surinam district, collected by D. C. Geijskes. Recent Paratypes (four lots): RMNH.MOL.55795, one left valve from Suriname, collected by D. C. Geijskes; RMNH.MOL.55792, a right and a left valve (Fig. 152 a, b) from Coppename Punt, Saramacca district, collected by D. C. Geijskes; RMNH.MOL unnumbered, a small right valve from the beach near Popogaimama Creek, Saramacca district, collected by D. C. Geijskes; USNM 638959, one young right valve from Coquette station 2, 6°23′N, 55° 05.5′W, 15 fathoms, 5/11/1957 (mentioned in the original description but not found).
Geographic distribution: Caribbean Sea to northern Brazil.
Stratigraphic distribution: Recent.
Synonyms: *Micromactra miskito* Petuch, 1998: 40.
Remarks: The type material of *Standella (Micromactra) surinamensis* is closely related to *S. (M.) janeiroensis*. Besides, Altena (1968) mentioned similarities with the Miocene species *Mactra californica maracaibensis* H. K. Hodson (in Hodson and Hodson, 1931: 20) from Venezuela. The revision of Hodson types (PRI 24052, holotype) revealed a shell with stronger umbonal undulations and

a more defined keel delimiting the escutcheon as suggested by Altena (1968). However, Altena (1968) admitted that the differences between *S. (M.) surinamensis* and the two mentioned species are small and future comparison must be done. Recently, Huber (2015) suggested that these similarities are enough to separate the northern *surinamensis* from the southern *janeiroensis*. At the moment, this point of view is accepted until genetic data reveal the presence of only one species with anti-tropical distribution. Finally, the type material of *Micromactra miskito* (CMNH 47364, holotype; CMNH 47352, paratype), described from Puerto Cabezas, Miskito coast, Nicaragua, confirms it as a junior synonym of Altena species.

Relevant literature

Taxonomy: Altena, 1968: 172–174; Huber, 2015: 834; Palacio et al., 2010: 406; Petuch, 1998: 40; Signorelli and Pastorino, 2012a: 93–94; von Cosel, 1973: 52.

Subgenus Mactroderma *Dall, 1894d*

Similar to *Standella*, shell more trigonal-ovate, less elongate, pallial sinus broad, posterior gap larger.

Type species: *Mactra velata* Philippi, 1849, by original designation.

Geographic distribution: West coast of central and northern South America.

Remarks: The taxonomic placement of this subgenus must be also genetically checked. It shares morphological characters and geographical distribution with *Standella*. However, the revision of type species and additional material, revealed differences in shell shape and gap pallial sinus morphology. At the moment, *Mactroderma* is herein understood as a subgenus of *Standella*. In this subgenus two species are recognized as living in American waters. They are *Standella (Mactroderma) isthmica* (Pilsbry and Lowe, 1932) and *Standella (Mactroderma) velata* (Philippi, 1849).

Standella (Mactroderma) isthmica **(Pilsbry and Lowe, 1932) (Fig. 3.9a–d)**

Shell sub-ovate to sub-trigonal, equilateral, moderately posteriorly elongated, relatively thick; medium size to large, length to 60 mm; umbo broad with commarginal undulations; anterodorsal margin straight, anterior end rounded; posterior end broadly truncate; posterodorsal area moderated, defined by a low ridge from umbo to posterior end, broadly gaping at the posterior end; internally white, porcelaneous; hinge plate concentered in the cardinal area, with strong, but short lateral teeth; internal ligament placed into chondrophore slightly ventrally projected; pallial sinus broad, variable in depth; muscle scar subcircular, strong, the posterior one slightly larger.

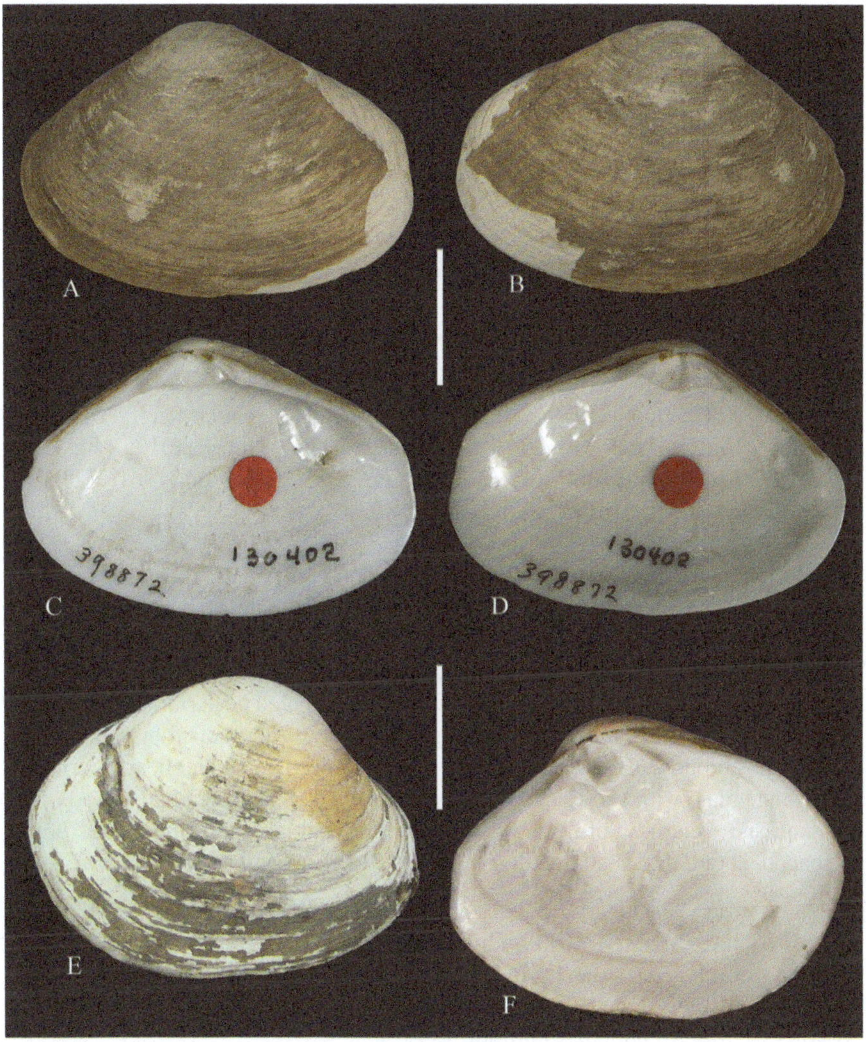

Fig. 3.9 (**a–d**) *Standella (Mactroderma) isthmica* (Pilsbry and Olsson, 1932), ANSP 130402, holotype. (**e**, **f**) *Standella (Mactroderma) velata* (Philippi, 1849), MNHNS 101194, holotype, a single right valve. Scale bar: (**a–d**) 1 cm, (**e**, **f**) 2 cm

TYPE LOCALITY: Near Panama City.

TYPE MATERIAL: ANSP 130402, Holotype; ANSP 398872 (ex 130402), one paratype.

GEOGRAPHIC DISTRIBUTION: West coast of central and northern South America from Champerico, Retalhuleu, Guatemala to Callao, Peru and Galápagos Islands, Ecuador.

STRATIGRAPHIC DISTRIBUTION: Recent.

SYNONYMS: *Mactra (Micromactra) vanattae* Pilsbry and Lowe, 1932: 90; *Mactra (Micromactra) vanattae acymata* Pilsbry and Lowe, 1932: 90; *Mactra (Micromactra) fonsecana* Hertlein and Strong, 1950: 232.

REMARKS: This species was recently revised by Coan and Valentich-Scott (2012: 549). Type material examination confirmed three junior synonyms of *S. (M.) isthmica*. They are *Mactra (Micromactra) vanattae* (ANSP 155921, syntype), *Mactra (Micromactra) vanattae acymata* (ANSP 152560, 2 syntypes) and *Mactra (Micromactru) fonsecana* (CAS 61534, holotype; CAS 61535, 64511 and 159007, paratypes). The first two junior synonyms were collected at Panama City, whereas *Mactra (Micromactra) fonsecana* was collected in Potosi and Monypenny Point, Gulf of Fonseca, Nicaragua.

RELEVANT LITERATURE

Taxonomy: Cauquoin, 1969a: 1023–1024; Coan and Valentich-Scott, 2012: 549–550; Hertlein and Strong, 1950: 232, 1955: 204; Huber, 2010: 444; Keen, 1971: 204; López de Mesa and Cantera, 2015: 8; Olsson, 1961: 324–325; Paredes et al., 2016: 141; Zilch, 1954: 86.

Standella (Mactroderma) velata (Philippi, 1849) (Fig. 3.9e, f)

Shell subcircular to ovate, inequilateral, thick, large, length to 120 mm; external surface with irregular and fine commarginal ribs; elongated posteriorly; anterior and posterior end rounded, anterodorsal margin straight; posterodorsal margin slightly convex; posterodorsal slope set off by a low ridge; brownish and heavy periostracum; hinge plate with short lateral teeth and cardinal teeth flanked posteriorly by a trigonal and ventrally projected chondrophore; pallial sinus moderate, broad, U-shaped, about 35% shell length.

TYPE LOCALITY: "Panama".

TYPE MATERIAL: MNHNS 101194, holotype, a single right valve.

GEOGRAPHIC DISTRIBUTION: From Isla Venado, Panama to Tacna, Peru.

STRATIGRAPHIC DISTRIBUTION: Pleistocene—Recent.

SYNONYMS: *Mactra paitensis* Philippi, 1893: 12.

REMARKS: This species was described twice by Philippi. First, in 1849 after specimens collected at Panama. And later, in 1894 as *Mactra paitensis* (MNHNS 228, syntype) from specimens collected at Paita, Peru. Coan and Valentich-Scott (2012) mentioned that records of this species further north of its range are erroneous.

RELEVANT LITERATURE

Taxonomy: Berú et al., 2014; Carpenter in Lea et al., 1860: 2; Cauquoin, 1969a: 1019–1020; Coan and Valentich-Scott, 2012: 550; Huber, 2010: 444; Keen, 1971: 202; Olsson, 1961: 322–323; Paredes et al., 2016: 141; Poutiers, 1995: 144; Ramírez et al., 2003: 268; Zilch, 1954: 86.

Genus *Trinitasia* Maury, 1928

Shell trigonal to subcircular, anterodorsal area concave; pallial sinus deep, V-shaped; pallial line evidently separated from ventral edge; lateral teeth short, close to cardinals, accessory lamella strong; anterior end round and high.

TYPE SPECIES: *Thyasira sanctiandreae* Maury, 1925, by original designation.
GEOGRAPHIC DISTRIBUTION: Atlantic coasts of Central and South America.
REMARKS: The genus *Trinitasia* was described by Maury (1928: 318) to include the Miocene species *Thyasira sanctiandreae* Maury, 1925 recorded from deposits of Trinidad. It was originally placed in the family Thyasiridae on the advice of Dall, as mentioned by Maury (1925). Subsequent authors had placed this genus in Lucinidae (Chavan in Moore, 1969; Vokes, 1980). However, based on internal shell features Woodring (1982) showed that *Trinitasia* belongs to Mactridae. Maury's genus was exclusively mentioned for fossil species (Maury, 1928, Woodring, 1982) still Signorelli and Pastorino (2012b: 32–35) relocated the Recent *Mactrella iheringi* Dall, 1897 into *Trinitasia*. The Recent species was mentioned as belonging to different genera (e.g. *Mactra*, *Mactrella*, *Mactrellona*) in the literature (Dall 1897; Altena, 1971; Rios, 1975, 1985, 1994; among others). However, shell characters confirmed the new combination proposed by Signorelli and Pastorino (2012b).

Trinitasia iheringi *(Dall, 1897) (Fig. 3.10a–b)*

Shell trigonal to subcircular, inequilateral, strong, umbones prosogyrate, inflated; dorsal margin concave in front of umbones, anterior end rounded; escutcheon not defined, ventral margin convex; external surface smooth, maximum length measured 57 mm (MACN-In 1817); internally white, left hinge with two short lateral teeth (AII and PII), with one cusp, cardinal tooth V-shaped composed of two simple divergent cardinals (2a and 2b) flanked by accessory lamella (4b); right hinge with two anterior lateral teeth (AI and AIII) elongated and similar in size and shape, two posterior lateral teeth (PI and PIII), the ventral one larger and elongated; with two cardinal teeth (3a and 3b) unfused and fragile; a trigonal chondrophore ventrally developed completes the hinge plate; pallial sinus V-shaped and deep to about half shell length; ultrastructure of the shell with an outer crossed lamellar layer of elongated crystallites and an inner layer with complex crossed lamellar structure; siphons entirely fused and covered by the periostracum, internal papillae observed along the length axis; body laterally compressed with a rounded and large foot; ctenidia composed of two elongated and lenticular demibranchs, inner demibranch larger with a food groove over the ventral edge, not observed in the outer one; labial palps trigonal, moderately elongated in relation to ctenidia size.

Fig. 3.10 (**a**, **b**) *Trinitasia iheringi* (Dall, 1897), USNM 107632, holotype; (**c**, **d**) *Tumbeziconcha goniocyma* (Pilsbry and Olsson, 1932), ANSP 155567, holotype; (**e**–**h**) *Tumbeziconcha thracioides* (Adams and Reeve, 1850), NHMUK 1874.12.11.393, holotype. Scale bar (**a**, **b**) 5 cm, (**c**, **d**) 5 mm, (**e**–**h**) 2 cm

TYPE LOCALITY: San Pablo, Brazil. Collected by H. von Ihering.
TYPE MATERIAL: USNM 107632, holotype, one left valve without periostracum.
GEOGRAPHIC DISTRIBUTION: From Florida, United States to Santa Catarina, Brazil.
STRATIGRAPHIC DISTRIBUTION: Recent.
SYNONYMS: *Mactra kempfi* Cauquoin, 1969b: 1175; *Mactra inceri* Petuch, 1998: 39.
REMARKS: The examination of types of *Mulinia kempfi* Cauquoin, 1969b (MNHN-IM-2000-34898, syntype) and *Mactra inceri* Petuch, 1998 (CMNH 47338, holotype, single right valve; CMNH 47346, 47350, paratypes) confirmed both nominal species as junior synonyms of *Trinitasia iheringi*. The former was collected off Amazonas mouth (Station 1774 of R/V Calypso), whereas the latter was collected at Puerto Cabezas, Miskito Coast, Nicaragua. *Trinitasia iheringi* constitutes an evidence of a Miocene genus, living in American waters.
RELEVANT LITERATURE

Anatomy and Morphology: Signorelli and Pastorino, 2012b.
Paleontology: Maury, 1920, 1925: 166, 1928; 318.
Taxonomy: Altena, 1971: 54; Cauquoin, 1969b: 1175; Dall, 1897:123, 1902: 510, 1915: 62; Diaz Merlano and Puyana Hegedus, 1994: 85; Palacio et al., 2010: 406; Petuch, 1998: 39; Rios, 1970: 196, 1975: 236, 1985: 243, 1994: 265, 2009: 536; Signorelli and Pastorino, 2012b: 32–35.

Genus *Tumbeziconcha* Pilsbry and Olsson, 1935

Shell fragile, with commarginal to oblique undulations on the external surface, observed internally; chondrophore small, pallial sinus deep and broad.

TYPE SPECIES: *Mactra thracioides* Adams and Reeve, 1850, by original designation.
GEOGRAPHIC DISTRIBUTION: Eastern Pacific Ocean from Mexico to Peru.
REMARKS: Two Recent species distributed along the eastern Pacific are included in the genus *Tumbeziconcha*. They are the type species *T. thracioides* (Adams and Reeve, 1850) and *T. goniocyma* (Pilsbry and Lowe, 1932), recently included in *Tumbeziconcha* by Huber (2015). This opinion is shared. However, the inclusion of the Indo-Pacific *Mactra plicatilis* Reeve, 1854a and the West African *Raeta senegalica* Von Cosel, 1995 in *Tumbeziconcha* is not definitive. Although, all species show commarginal undulating sculpture, true *Tumbeziconcha* show commarginal undulations with a defined angle in the central region of the shell. In addition, a more inflate umbo and less elongated shell shape distinguish American species. Definitely, additional analysis will confirm the generic position of non-American species.

Tumbeziconcha goniocyma *(Pilsbry and Lowe, 1932)* *(Fig. 3.10c–d)*

Shell ovate, thin, fragile, translucent, ovate to triangular, moderately inflated; umbo small placed at half shell length; anterior end broadly rounded, posterior end triangular; external surface with commarginal and regular undulations; very fine concentric striation crosses the undulations obliquely; posterodorsal area set off by raised keel; lunular area smooth, lanceolate; periostracum thin, translucent to light; hinge plate with the anterior lateral teeth very close placed to the V-shaped cardinal tooth, posterior lateral more elongated than anterior, chondrophore not ventrally projected; pallial sinus not easily visible due to the external ornamentation.

TYPE LOCALITY: San Juan del Sur, Nicaragua.
TYPE MATERIAL: ANSP 155567, holotype one right valve.
GEOGRAPHIC DISTRIBUTION: from Mazatlán, Mexico to Colombia.
STRATIGRAPHIC DISTRIBUTION: Recent.
SYNONYMS: None.
REMARKS: The taxonomic position of this species is not conclusive. Huber (2010) excluded it from the Indo-Pacific *Mactrinula*, based on morphological differences of umbo, size and sculpture. However, Coan and Valentich-Scott (2012) based on the posterior slope defined by a keel included it into *Mactrinula*. Hinge plate of *goniocyma* show lateral teeth slightly larger compared to the type species *T. thracioides*. But differences are more evident in comparison with *M. plicataria*, type species of *Mactrinula*. Huber opinion is, at the moment, accepted.
RELEVANT LITERATURE

Taxonomy: Bernard, 1983: 39; Hertlein and Strong, 1950: 234–235, 1955: 205; Keen, 1971: 202; Olsson, 1961: 330; Pilsbry and Lowe, 1932: 90; Coan and Valentich-Scott, 2012: 547.

Tunmbeziconcha thracioides *(Adams and Reeve, 1850)* *(Fig. 3.10e–h)*

Shell ovate to subcircular, thin, inflated, short, shell length from 30 to 45 mm, umbo placed at half shell length, prosogyrate; anterior end rounded; posterior end sharply rounded, narrower than the anterior, posterodorsal area straight, external surface with commarginal undulations on the anterior slope turning to oblique to the middle and posterior slopes; with fine concentric wrinkles; without keel; light brown periostracum, thin; internally white, hinge plate narrow; cardinal teeth thin flanked posteriorly by a ventrally projected chondrophore, lateral teeth short; external ligament separated from internal by a ridge; pallial sinus broad and deep, about 60% shell length, posteriorly gaped.

Type locality: "Eastern Seas" erroneously registered. Probably obtained from some station in the Panamic-Pacific region visited by the Samarang expedition.

Type material: NHMUK 1874.12.11.393, holotype.

Geographic distribution: West coast of Central America and South America from El Salvador to northern Peru.

Stratigraphic distribution: Recent.

Synonyms: None.

Remarks: *Tumbeziconcha thracioides* has been mentioned as living in the west coast of Central and northern South America. It can be distinguished from *T. goniocyma* by the absence of a keel in the posterodorsal area.

Relevant literature

Taxonomy: Bernard, 1983: 39; Coan and Valentich-Scott, 2012: 557; Conrad, 1868: 46; Keen, 1971: 204; Lamy, 1918; 386; Olsson, 1961: 333–334; Paredes et al., 2016: 141; Pilsbry and Olsson, 1935: 119–121; Ramírez et al., 2003: 268; Reeve, 1854a: sp. 116; Zilch, 1954: 86.

Chapter 4
The Subfamilies Lutrariinae and Darininae (Mollusca: Bivalvia)

Abstract In this chapter, the species included in the subfamily Lutrariinae Gray, 1853 and Darininae Signorelli in Carter et al., 2011 represented in American marine waters are redescribed. In relation to Lutrariinae, *Tresus* is the only genus of the subfamily living in American waters. This genus includes three Pacific species: *T. allomyax* Coan and Valentich-Scott, 2000, *T. capax* (Gould, 1850) and *T. nuttallii* (Conrad, 1837). The subfamily Darininae is represented by the genus *Darina* Gray, 1853 in the southern tip of the American continent. This genus includes two living species: *D. solenoides* (King, 1832) and *D. rustica* Huber, 2010.

Keywords Lutrariinae · *Tresus* · Darininae · Darina

Subfamily Lutrariinae Gray, 1853

[nom. correct. H. Adams and A. Adams, 1856, pro subfamily Lutrariana [sic] Gray, 1853, p. 42; = Zenatiinae Dall, 1895b: 209, 212; = Resaniinae Marwick, 1931, p. 73, no description, but made available by its subsequent use as a valid name before 2000 by Beu, 1966, p. 68]

Shell large, inequilateral, elongated, widely gaping, external surface usually smooth with commarginal lines; periostracum thick, posteriorly extended over the siphons; hinge plate with large chondrophore, lateral teeth short, close to the cardinals; pallial sinus deep.

Genus Tresus *Gray, 1853*

[not preoccupied by *Tresus* Walckenaer, 1833, p. 438, vernacular and nomen nudum; = *Cryptodon* Conrad, 1837, p. 235 (*non* Turton, 1822, p. 121; objective); = *Schizothaerus* Conrad, 1853, p. 199 (objective.)]

Shell ovate, moderately elongated, posteriorly gaped, truncated, anterior end rounded; external surface smooth; external and internal ligament separated by a

J. H. Signorelli, *The Superfamily Mactroidea (Mollusca:Bivalvia) in American Waters*, https://doi.org/10.1007/978-3-030-29097-9_4

shelly plate; chondrophore ventrally projected, strong; lateral teeth very short, pallial sinus deep and broad.

TYPE SPECIES: *Lutraria maxima* Middendorff, 1849 (*non* Jonas, 1844) (= *L. capax* Gould, 1850), by monotypy.
GEOGRAPHIC DISTRIBUTION: West North America, East Asia.
REMARKS: the genus *Tresus* currently groups four species, three of them living in the Pacific coast of North America. They are *Tresus allomyax* Coan and Valentich-Scott, 2000, *T. capax* (Gould, 1850) and *T. nuttallii* (Conrad, 1837). The generic names *Tresus* and *Schizothaerus* were proposed almost simultaneously, however *Tresus* has priority (Coan et al., 2000). Hinge plate morphology and shell shape widely gaping place this genus in Lutrariinae.

***Tresus allomyax* Coan and Valentich-Scott, 2000 (Fig. 4.1a–d)**

Shell ovate, thick, subequilateral, large, length to 150 mm; umbos low and not inflated, placed near anterior third of shell; external surface smooth with irregular growth lines, brownish periostracum; anterior end rounded; posterior end weakly truncate, with a short and narrow siphonal gap; internally white, porcelaneous; hinge plate deeply grooved on both sides of umbo, with conspicuous dorsal flanges; chondrophore ventrally projected; pallial sinus broad and deep.

TYPE LOCALITY: off Bodega Head, Sonoma county, California (38.3° N, 123.1° W), 73 m depth.
TYPE MATERIAL: CAS 093991, holotype, articulate specimen. CAS 099254, paratype from type localities. CAS 093994 and 093995, additional paratypes from off Clam beach, and from off Orick, respectively, both in Humboldt county, California.
GEOGRAPHIC DISTRIBUTION: West coast of United States, from Redding Rock, Humboldt county to Sonoma County, California (Coan et al., 2000).
STRATIGRAPHIC DISTRIBUTION: Recent.
SYNONYMS: *Tresus pajaroanus* of authors (*non* Conrad, 1857: 192).
REMARKS: *Tresus allomyax* was named after Dinnel and DeMartini (1974) who recognized a third species of the genus living along the coast of northern California. It was incorrectly referred to as *Tresus pajaroanus*, a fossil taxon recorded in Cenozoic formations of northwest California and Oregon (Roth, 1979). Shell outline clearly separates *T. allomyax* from the more elongated *T. nuttallii* and the more rhomboidal *T. capax* (See Coan et al., 2000: 460 for further analysis).
RELEVANT LITERATURE

Taxonomy: Coan and Valentich-Scott in Carlton, 2007: 848; Coan and Valentich-Scott, 1997: 17; Coan et al., 2000: 460–463; Dinnel and DeMartini, 1974: 44–46; Holm, 2008.

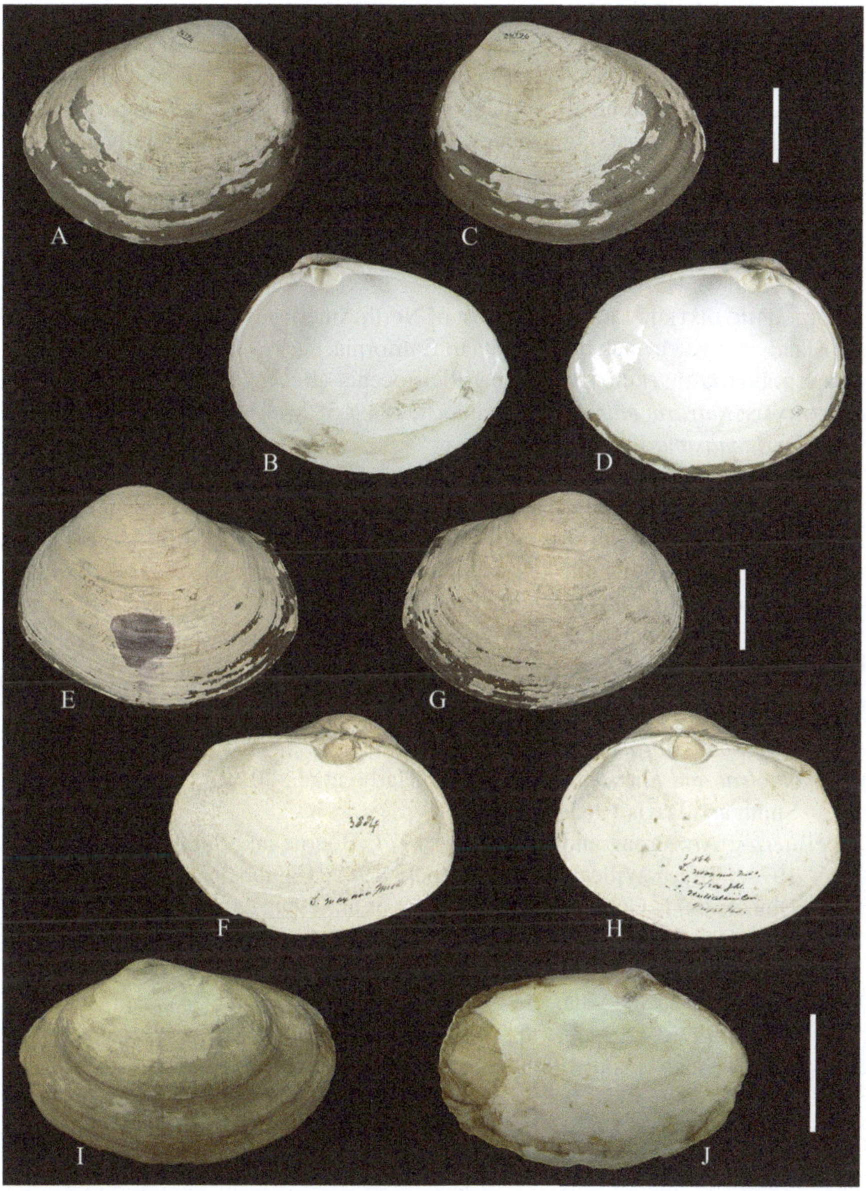

Fig. 4.1 (**a–d**) *Tresus allomyax* Coan and Valentich Scott, 2000, CAS 093991, holotype; (**e–h**) *Tresus capax* (Gould, 1850), USNM 3884, holotype; (**i, j**) *Tresus nuttallii* (Conrad, 1837), NHMUK 1861.5.20.134, holotype. Scale bar (**a–h**) 4 cm, (**i, j**) 2 cm

Tresus capax (Gould, 1850) (Fig. 4.1e–h)

Shell ovate to rhomboidal, large, thick, length to 280 mm, inflate, anterior end rounded, posterior end nearly vertical truncate, posteriorly broadly gaped; ventral margin distinctly convex; hinge plate with a large chondrophore ventrally projected and small cardinal and short lateral teeth; pallial sinus broad and deep; usually with commensal crabs belonging to the genus *Pinnixa* living into the mantle cavity.

TYPE LOCALITY: Puget Sound, Washington state, United States.

TYPE MATERIAL: USNM 3884, holotype.

GEOGRAPHIC DISTRIBUTION: West coast of North America from Northern Shumagin islands and Kachemak bay, Alaska to California.

STRATIGRAPHIC DISTRIBUTION: Upper Pleistocene—Recent.

SYNONYMS: *Lutraria maxima* Middendorff, 1849: 582 (invalid: junior homonym of *Lutraria maxima* Jonas, 1844: 34).

REMARKS: *Tresus capax* constitutes the northern species of the genus in American waters. It was originally included into *Lutraria* by Gould (1850) and quickly synonymized with *Lutraria maxima* Middendorff, 1849 (ZISP 213/1, collected at Sitcha, ZISP 214/2 collected at Columbia River, syntypes) in the U. S. Explorer Expedition Mollusca volume (Gould, 1852: 395). However, Middendorff's name is a junior homonym of *Lutraria maxima* Jonas, 1844. *Tresus capax* was widely studied due to fisheries purposes (see Relevant Literature section).

RELEVANT LITERATURE

Anatomy and Morphology: Kellogg, 1915.

Ecological and Environmental studies: Harbo et al., 2012; Pearce, 1965, 1966; Smith and Davis 1965.

Fisheries: Armstrong and Armstrong, 1974; Bourne and Smith, 1972a, 1972b, 1972c; Breed-Willecke and Hancock, 1980; Campbell and Bourne, 2001; Dinnel, 1971; Goodger, 1972; Lauzier et al., 1998; Machell and DeMartini, 1971; Zhang and Campbell, 2002.

Paleontology: Addicott, 1963; Grant and Gale, 1931.

Taxonomy: Coan and Valentich-Scott in Carlton, 2007: 848; Coan and Valentich-Scott, 1997: 17; Coan et al., 2000: 461, 463; Dall, 1916a: 40; Frizzell, 1930; Haderlie and Abbott, 1980; Huber, 2010: 452; Reid and Rauchert, 1976; Reid, 1969; Robinson and Breese, 1982; Swan and Finucane 1952; Wendell et al., 1976.

Tresus nuttallii (Conrad, 1837) (Fig. 4.1i, j)

Shell ovate-elongate, large, thick, length to 200 mm, umbo prominent; longer posteriorly, posterior end obliquely truncate; broadly gaped; anterior end rounded; external surface smooth with irregular commarginal striae; brownish periostracum; hinge plate with posterior cardinal very close to the large and ventrally projected chondrophore, posterior lateral teeth short, anterior lateral teeth absent; pallial sinus deep and broad.

Type locality: Santa Barbara, California, United States.

Type material: NHMUK 1861.5.20.134, holotype, a juvenile and eroded single left valve.

Geographic distribution: West coast of North America from Alaska to Baja California, México.

Stratigraphic distribution: Miocene—Recent.

Synonyms: *Lutraria inflata* Dunker, 1853: 112; *Spisula longa* Dall, 1921b: 22; *Schizothaerus nuttallii bighopensis* Henderson, 1931: 33.

Remarks: The Recent *Lutraria inflata* (types not examined) described from the coast of California has been largely mentioned as a junior synonym of *T. nutallii* (Coan and Valentich-Scott, 2012 and literature therein). In addition, two fossil species are currently considered synonym of *T. nutallii*. They are *Spisula longa* (USNM 333116, syntypes) from Pleistocene deposits of San Quentin bay, and *Schizothaerus nuttallii bighopensis* (types not found) from Pleistocene deposits of Puget Sound. *Lutraria maxima* Jonas, 1844 was mentioned as a junior synonym of *T. nuttallii* (Bernard, 1983; Coan et al., 2000). However, this is a valid species distributed along the coast of southern China (Bernard et al., 1993; Huber, 2010; Coan and Valentich-Scott, 2012).

Relevant literature

Anatomy and Morphology: Des Voigne et al., 1970; Pohlo, 1964; Taylor and Smith, 1966; Patterson, 1933: 77–83.

Ecological and Environmental studies: Pearce, 1965; Smith and Davis, 1965; Stout, 1970; Peterson, 1977; Peterson and Andre, 1980.

Fisheries: Campbell et al., 1990, 2009; Campbell and Bourne, 2001; Clark et al., 1975; Clark, 1973; Frey, 1971: 27–29; Katkansky et al., 1969; Laurent, 1971; Lauzier et al., 1998; Stearns, 1891; Strathmann, 1987: 334–335.

Paleontology: Addicott, 1963; Grant and Gale, 1931: 404–405; Hertlein and Grant, 1972: 318; Valentine, 1957.

Taxonomy: Abbott and Hilgard, 1987:198–199; Coan and Valentich-Scott in Carlton, 2007: 848; Coan and Valentich-Scott, 1997: 17; Coan and Valentich-Scott, 2012: 558; Coan et al., 2000: 463; Dall, 1916a: 40; Haderlie and Abbott, 1980: 378; Henderson, 1931; Huber, 2010: 452; Keen, 1966: 170; MacGinitie, 1941; Strong, 1925; Swan and Finucane 1952.

Subfamily Darininae Signorelli in Carter et al., 2011

Shell thin, fragile, oval to subcircular, elongate, anteriorly and posteriorly gaped; umbo not inflated, centrally positioned; external ligament rudimentary, internal ligament place in a ventrally to posteroventrally strongly projected chondrophore; Hinge plate concentrated on the central area, with short lateral teeth; pallial sinus deep.

Genus Darina *Gray, 1853*

Shell very thin, fragile, elongated, compressed laterally; gaped at both ends; external ligament rudimentary, lateral teeth weak placed very close to the cardinals; pallial sinus deep about one-third of shell length.

TYPE SPECIES: *Erycina solenoides* King, 1832, by monotypy.

GEOGRAPHIC DISTRIBUTION: From Puerto Montt, Chile in the Pacific Ocean to Bahía Blanca, Argentina, in the Atlantic.

REMARKS: The genus *Darina* currently groups two species. *Darina solenoides* (King, 1832) and *D. rustica* Huber, 2010. Both species are herein considered as valid, however, they must be genetically checked.

Darina solenoides (King, 1832) (Fig. 4.2a–f)

Shell elliptical, fragile, inequilateral, young specimens pellucid; with a maximum length of about to 70 mm; small umbos placed behind midline of shell, not inflated, weakly prosogyrous; external surface smooth, with irregular, concentric growth lines only; anterior dorsal margin straight, larger than posterior margin, which is curved; posterior end with siphonal gap larger than anterior, both ends rounded; ventral margin almost straight, curved at distal portions; lunule and escutcheon poorly defined; internally white, hinge with one short, fragile anterior lateral tooth in left valve (AII), two fused cardinal teeth (2a and 2b) forming V-shaped tooth, tooth 2b larger than 2a, flanked by a lamella (4b); posterior lateral tooth (PII) similar in size to anterior, very close to the cardinals; right valve with two anterior lateral teeth (AI and AIII) different in size, ventral (AI) more elongated and stronger; cardinal teeth (3a and 3b) unfused, 3a larger; both posterior lateral teeth (PI and PIII) similar in size and shape, short, very fragile; pallial sinus deep, approximately half of shell length; pallial line very well defined separated from ventral margin. Anterior adductor muscle scar ellipsoidal, larger than more subcircular posterior; internal ligament well developed in very large chondrophore, ventrally projecting; external ligament rudimentary, placed behind the umbo; ultrastructure of the shell with two layers: outer with single cross lamellar structure and inner with complex cross lamellar structure; siphons unfused, only united at base very close to the siphonal retractor muscle; incurrent and excurrent siphons with very well-developed tentacles at apertures, all equal in size and position; siphonal papillae present longitudinally; external demibranch trigonal, not elongated, very high; internal demibranch elongated in anteroposterior axis; internal surface of external demibranch plicate and covered by cilia to facilitate food current; this demibranch unfused at any point with labial palps; labial palps small, trigonal, elongated, with plicate internal surfaces; outer labial palp fused to mantle, inner fused internal demibranch; posterior portion of mantle completely arched to the inside, leaving siphon naked.

TYPE LOCALITY: Puerto El Hambre in the Magellanic Strait.

TYPE MATERIAL: NHMUK 1837.12.1.879/884, 1968507 and 1859.9.19.59, nine syntypes six broken shells.

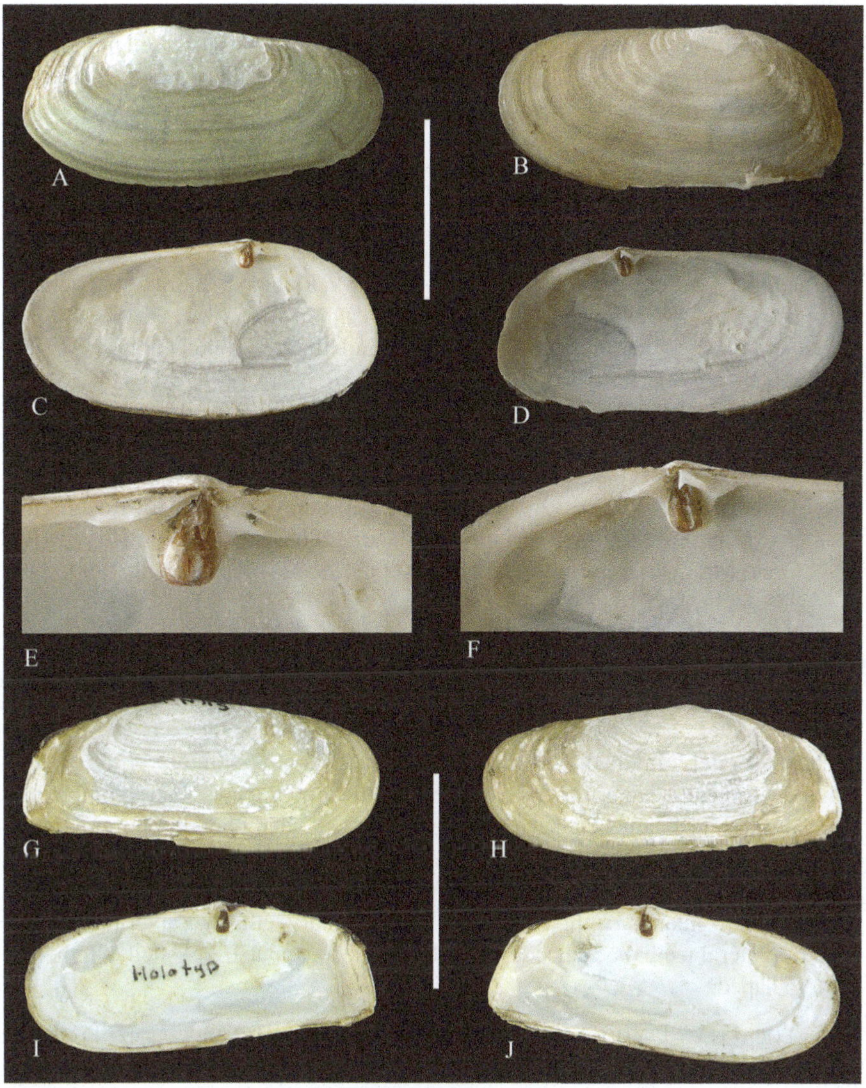

Fig. 4.2 (**a–f**) *Darina solenoides* (King, 1832), (**a**) *Lutraria tenuis* Philippi, 1845, ZMB, holotype, (**b–d**) paratypes of *L. tenuis*; (**e**, **f**) hinge plate detail; (**g–j**) *Darina rustica* Huber, 2010, NHMUK 20100578, Holotype. Scale bar (**a–j**) 2 cm

Geographic distribution: Punta Alta, Argentina, to Puerto Montt, Chile.

Stratigraphic distribution: Pleistocene—Recent.

Synonyms: *Darina declivis* Carpenter, 1864: 607, 637; *Lutraria kingi* Fischer, 1887: 1119; *Lutraria tenuis* Philippi, 1845: 50.

Remarks: *Lutraria tenuis* (ZMB 112740, syntypes), described from the Magellan Strait and *Darina declivis* (Redpath Museum 101, holotype), erroneously described from Vancouver, are confirmed as synonyms of *D. solenoides*. *Lutraria kingi* was proposed by Fischer (1887, 1119) to remove the supposed homonymy with *Lutraria solenoides* Lamarck, 1818, as he considered the genus *Darina* as a section of *Lutraria*. However, both genera are clearly distinct and valid.

RELEVANT LITERATURE

Anatomy and Morphology: Signorelli and Pastorino, 2011.

Ecological and Environmental studies: D'Amico et al., 2004; Ferrari et al., 2010; Gallardo and Lizarralde, 2014; Musmeci et al., 2013.

Paleontology: Ihering, 1907: 458; Nielsen and Valdovinos, 2008: 206; Richards and Craig, 1963: 137.

Parasitology: Cremonte, 2001; Cremonte and Ituarte, 2003; Gilardoni et al., 2013.

Taxonomy: Adams and Adams, 1856: 381; Bernard, 1983: 41; Carcelles and Williamson, 1951: 344; Carcelles, 1944: 284; Castellanos, 1970: 239; Chenu, 1862: 60; Coan et al., 2000: 451; Conrad, 1868: 46; Dall, 1895b: 212; Hanley, 1843 (in 1842–1856): 39; Huber, 2010: 453; Lamy, 1912a: 315; Rochebrune and Mabille, 1889: 108; Martínez y Saez, 1869: 14; Pastorino, 1995: 14; Gray, 1853: 42; Signorelli and Pastorino, 2011: 61–64; Stempell, 1899: 238–239; Zelaya, 2016: 252.

Darina rustica Huber, 2010 (Fig. 4.2g–j)

Shell moderately inflated, rhomboidal; ventral margin straight; posterior end obliquely truncate, anterior end rounded; both ends gaped; central umbo, not inflated; internal ligament placed in a vertically extended chondrophore; external surface smooth with a brownish yellow periostracum; hinge plate similar to *D. solenoides*, concentrated on cardinal area, left valve with an inverted V-shaped cardinal and two small laterals, right valve with two cardinal teeth, two anterior and two posterior lateral teeth on each side; pallial sinus deep, about 1/2 of shell length, narrow, V-shaped.

TYPE LOCALITY: Puerto Santa Cruz (50.0°S, 68.5°W) Santa Cruz Province, Patagonia, Argentina; at extreme low tide attached to algae; also, in shallow water, attached to algae, collected in 2002 by T. Narosky and A. F. Suarez.

TYPE MATERIAL: NHMUK 20100578, holotype; Paratypes 1 (30.1 mm, RA115) and 2 (32.2 mm, RA115 1) in collection M. Huber; Paratypes 3 (42 mm) and 4 (33 mm) in collection A. F. Suarez; Paratypes 5 (35 mm) and 6 (34 mm) in collection T. Narosky.

GEOGRAPHIC DISTRIBUTION: only known from the type locality and its surroundings.

STRATIGRAPHIC DISTRIBUTION: Recent.

SYNONYMS: None.

Remarks: The types of *Darina rustica* show a more inflated and thicker shell, with a rhomboidal shape and an enlarged and narrow pallial sinus. Huber (2010) was based on these morphological characters to distinguish his species from the well known *D. solenoides*.

Relevant literature

Taxonomy: Huber, 2010: 453.

Chapter 5
The Family Anatinellidae (Mollusca: Bivalvia)

Abstract The living species included in the family Anatinellidae Deshayes, 1853 distributed along the American continent are mentioned in this chapter. All American living species belong to the subfamily Kymatoxinae Stenzel and Krause, 1957. Two genera are recognized. The genus *Anatina* Schumacher, 1817 groups the species *A. anatina* (Spengler, 1802) and *A. cyprinus* (Wood, 1828). And the genus *Raeta* Gray, 1853 includes *R.* (Raeta) *plicatella* (Lamarck, 1818) and *R.* (Raeta) *undulata* (Gould, 1851).

Keywords Anatinellidae · Kymatoxinae · *Anatina* · *Raeta*

Family Anatinellidae Deshayes, 1853

[nom. correct. Thiele, 1934, p. 904 (in Thiele, 1929, 1931, 1934, 1935), pro Anatinellidae Deshayes, 1853: 43, name attributed by Gray to Deshayes] [*non* Spasskii and Spasskaya 1954, p. 61—Cestoda: Hymenolepidae]

Shell oblong-ovate inflated, posteriorly elongate, thin, fragile, slightly inequilateral, posteriorly gaping, umbos low, positioned slightly anterior of the midline, prosogyrate; sculpture with growth lines or commarginal striae and fine, radial lines; hinge plate very narrow with the chondrophore, slightly dorsally arched; two divergent cardinal teeth in the right valve (not fused) and a single, inverted V-shaped cardinal tooth in the left valve; lateral teeth; no true lateral teeth, but accessory shell lamella is observed in each valve; pallial line without sinus.

Subfamily Kymatoxinae Stenzel and Krause, 1957

[nom. correct. Kymatoxinae Stenzel and Krause, 1957:124, in Stenzel, Krause, and Twining, 1957] [=Pteropsinae Dall, 1894d (invalid); Pteropsellinae Keen, 1969, p. 605, not available from Keen, 1963: 92 [no description], proposed as a substitute name for Pteropsinae Dall, 1894e [also spelled Pteropsidinae by Dall], the latter

J. H. Signorelli, *The Superfamily Mactroidea (Mollusca:Bivalvia) in American Waters*, https://doi.org/10.1007/978-3-030-29097-9_5

invalid due to its type genus *Pteropsis* Conrad, 1860 is a junior homonym of *Pteropsis* Rafinesque, 1814 [Cnidaria]; Keen gave Pteropsellinae the priority of Pteropsinae, but Art. 40.2 does not apply; according to Art. 40.1, Kymatoxinae should not be replaced by Pteropsellinae on account of the synonymy of the type genera alone]

Shell thin, subequilateral, posterior gape narrow; thin periostracum; radially sculptured with striae, with posterodorsal keel; chondrophore ventrally projected; hinge plate with small cardinal teeth, lateral teeth absent.

Genus Anatina *Schumacher, 1817*

[=*Labiosa* Möller, 1832 (ex Schmidt MS) (nom. pro Anatina); *Cypricia* Gray, 1847; *Leucoparia* Mayer, 1867 (nom. Vam. Pro *Cypricia*)]

Shell ovate, very fragile, anterior end rounded; siphonal gap and defined by a line from umbones to posterior end; lunule and escutcheon evident; external ligament separated from resilium by lamella.

Type species: *Anatina pellucida* Schumacher, 1817 (= *Mactra anatina* Spengler, 1802), by monotypy.

Geographic distribution: Atlantic and Pacific coast of tropical America.

Remarks: *Anatina* Schumacher, 1817 (Mactridae) has priority over *Anatina* Lamarck, 1818 (Laternulidae). Some authors (e.g. Olsson, 1961; Vokes, 1968) considered *Anatina* Lamarck sensu Bosc (1816) as valid. However, Bosc cited only one species in vernacular. This nomenclatorial problem was resolved by Keen (1961). Within the genus *Anatina* of Schumacher, two Recent species are included. They are *Anatina anatina* (Spengler, 1802) and *A. cyprinus* (Wood, 1828). *Anatina cyprinus* (Wood, 1828) has been cited widely as a synonym of *A. anatina* (e.g. Dall, 1894b; Wilkins, 1957; Keen, 1958; Olsson, 1961). However, Spengler species is more subcircular, less elongated and has a smaller posterior gap. Other species have been described within the genus *Anatina* from other regions (*Anatina grayi* Huber, 2015 or *Anatina inconstans*, von Cosel, 1995). However, they were eventually relocated after morphological revisions (von Cosel and Gofas, 2018). The genus *Anatina* is an American genus.

Anatina anatina **(Spengler, 1802) (Fig. 5.1a–d)**

Shell subcircular, fragile, pellucid, inequilateral, large, length to 65 mm in length, umbones opisthogyrate and moderately inflated; anterodorsal margin straight, posterior very short; anterior end rounded, posterior area delimited by a line from the umbones to middle posterior end; siphonal gap formed by shell folds; external surface with concentric ridges more conspicuous along the umbonal area; internally white with a deep and high pallial sinus of about half of shell length; right hinge with two unfused cardinal teeth (3a and 3b), one anterior lateral tooth (AI) aligned with the 3a; one posterior lateral tooth (PI) short; left hinge with V-shaped cardinal

Fig. 5.1 (**a–d**) *Anatina anatina* (Spengler, 1802), ZMUC-BIV, syntypes; (**e**, **f**) *Anatina cyprinus* (Wood, 1828), NHMUK 2010.06.08, holotype; (**g**, **h**) *Raeta plicatella* (Lamarck, 1818), MNHN-IM-2000-5011, syntype; (**i–l**): *Raeta undulata* (Gould, 1851), MCZ 169453, holotype. Scale bar (**a–l**) 2 cm

tooth (2a and 2b), 2b more elongated, one anterior and one posterior lateral tooth (AII and PII) complete the hinge.

TYPE LOCALITY: "South America".

TYPE MATERIAL: ZMUC-BIV, two syntypes.

GEOGRAPHIC DISTRIBUTION: From Massachusetts, United States to Porto Belo, Santa Catarina, Brazil.

STRATIGRAPHIC DISTRIBUTION: Upper Miocene—Recent.

SYNONYMS: *Anatina pellucida* Schumacher, 1817: 126, *Lutraria lineata* Say, 1822: 310, *Mactra papyracea* Lamarck, 1818: 470 (*non* Gmelin, 1791), *Mactra recurva* Wood, 1828: 4 and *Mactra nuttallii* Reeve, 1854a: sp. 125 (*non* Conrad, 1837).

REMARKS: *Anatina anatina* is a valid mactrid species living along the western Atlantic Ocean. Type material and the examination of the original illustrations confirm that *Anatina pellucida* Schumacher (unnecessary substitute name for *Mactra anatina*), *Lutraria lineata* (ANSP 51313, syntypes), *Mactra papyracea* Lamarck, 1818 (*non* Gmelin, 1791), *Mactra recurva* (NHMUK 20100621, holotype) and *Mactra nuttallii* Reeve, 1854a (vol. 8, pl. 21, sp. 125) are junior synonyms of Spengler species. The Pacific species *Anatina cyprinus* has been mentioned as a junior synonym of *A. anatina* (see, e.g. Dall, 1894b; Wilkins, 1957; Keen, 1958; Olsson, 1961). However, it shows a more subcircular, less elongated and smaller posterior gap than the Atlantic *A. anatina*.

RELEVANT LITERATURE

Ecological and Environmental studies: Davies et al., 1989; Staff and Powell, 1999.

Taxonomy: Camp et al., 1998: 12; Conrad, 1868: 42; Dall, 1894c: 27; Keen, 1961; Lamy, 1918: 349; Mikkelsen and Bieler, 2007: 372; Rios, 1970: 197, 1975: 237, 1985: 244, 1994: 266, 2009: 538; Robinson and Montoya, 1987: 396; Signorelli and Pastorino, 2012b: 45–47; Signorelli and Carter, 2016; Tallarico et al., 2014: 101; Turgeon et al., 2009: 730.

Anatina cyprinus (Wood, 1828) (Fig. 5.1e, f)

Shell elongate-ovate, thin, translucent, inequilateral, equivalve, anteriorly moderately inflated; posteriorly gaped, posterior slope set of by a winged keel; external surface smooth with fine and irregular commarginal striae, more visible and regular on umbonal area; periostracum light brown to yellowish, thin; internally white; hinge plate narrow, with only one anterior and one posterior lateral tooth in both valves, placed very close to the cardinals; internal ligament placed on a very ventrally projected chondrophore; pallial sinus deep and broad.

TYPE LOCALITY: "Peru".

TYPE MATERIAL: NHMUK 2010.06.08, holotype.

GEOGRAPHIC DISTRIBUTION: West coast of North America, from Baja California, Mexico to northern Peru.

STRATIGRAPHIC DISTRIBUTION: Recent.

SYNONYMS: none.

REMARKS: This species has been widely cited as a synonym of *A. anatina* (see, e.g. Dall, 1894b; Wilkins, 1957; Keen, 1958; Olsson, 1961). All these authors considered the name *Anatina lineata* of Say (1822) valid for the Atlantic species. The examination of related type material revealed that *A. cyprinus* is the correct name for the Pacific species.

RELEVANT LITERATURE

Taxonomy: Wood, 1828: 4; Lamy, 1918: 349; Wilkins, 1957: 164; Keen, 1961, 1971: 206; Olsson, 1961: 333; Coan and Petit, 2011: 20; Coan and Valentich-Scott, 2012: 560; Poorman and Poorman, 1978: 371; Paredes et al., 2016: 141; Ramírez et al., 2003, 268; Signorelli and Carter, 2016.

Genus Raeta *Gray, 1853*

[= *Lovellia* Mayer, 1867: 286]

Shell trigonal, fragile, pellucid, resembling Anatina, but more convex; completely ornamented in the external surface by concentric wrinkles, posteriorly; anterior and posterior end rounded, lateral teeth rudimentary and close to the cardinals.

TYPE SPECIES: *Mactra campechensis* Gray, 1825 (= *Lutraria plicatella* Lamarck, 1818), by monotypy.

GEOGRAPHIC DISTRIBUTION: Europe, Pacific coast of Central and South America, western Atlantic and Asia.

REMARKS: The genus *Raeta* includes five Recent species widely distributed with several synonyms mentioned in the literature (Carcelles, 1944; Rios, 1994, 2009; Harry, 1969; Huber, 2010; Signorelli and Pastorino, 2012b). Currently, it includes *Raeta* (*Raeta*) *meridionalis* Tate, 1889 from Southern Australia; *Raeta* (*Raetina*) *pellicula* (Reeve 1854a) and *Raeta* (*Raetella*) *pulchella* (Adams and Reeve 1850) both from the Indo-Pacific region from India to Japan; *Raeta* (*Raeta*) *undulata* (Gould 1851) from eastern Pacific and *Raeta* (*Raeta*) *plicatella* (Lamarck 1818) from western Atlantic Ocean and eastern Pacific. The Atlantic *R. plicatella* was recently reported from Nayarit coast, Mexico (Skoglund, 2000; Coan and Valentich-Scott, 2012).

Subgenus Raeta

Raeta (*Raeta*) *plicatella* (Lamarck, 1818) (Fig. 5.1g, h)

Shell trigonal to ovoid, subequilateral, pellucid, inflated through the anterior portion, maximum length to 75 mm; external surface with diagnostic concentric strong wrinkles from umbones to ventral margin; anterodorsal and ventral margin rounded, posterior end rounded but more defined than the anterior; lunule and escutcheon not defined by a line or shelly ridge; interior white with soft wrinkles that reveal the external ornamentation; deep pallial sinus with a length of about 2/3 shell length; adductor muscle scar similar in size and shape, the anterior one more elongated dorsoventrally; right hinge with only one anterior and one posterior lateral tooth and two cardinal teeth not fused; laterals very short and close to the cardinals; left hinge with the inverted V-shaped cardinal teeth closely flanked by the accessory lamella; one anterior and one posterior lateral tooth; chondrophore very well developed; ultrastructure of the shell with the same layers that other mactrids; siphons completely fused, the inhalant bigger than the exhalant; two demibranchs at both sides of the body dorsally fused to the mantle with an elongated morphology; the inner demibranch bigger than the outer with a food grove in the ventral edge; labial palps notoriously developed, trigonal, placed surrounding the mouth and fused dorsally with the visceral mass.

TYPE LOCALITY: "l'Indian Ocean" is mentioned in the original description.

TYPE MATERIAL: MNHN-IM 2000-5011, two syntypes, one of them broken.

GEOGRAPHIC DISTRIBUTION: From Virginia, United States to Monte Hermoso, Buenos Aires province, Argentina, Pacific coast of America from California United States to Panama (Coan and Valentich-Scott, 2012: 558).

STRATIGRAPHIC DISTRIBUTION: Lower Pliocene—Recent.

SYNONYMS: *Lutraria canaliculata* Say, 1822: 311, *Mactra campechensis* Gray, 1825: 134 and *Raeta perspicua* Hutton, 1873: 65.

REMARKS: This species was recently mentioned as living in the west coast of Central America, from Baja California to Sonora, Mexico (Skoglund, 2000; Coan and Valentich-Scott, 2012). It was widely mentioned in the literature as living along the Atlantic coast of America from North Carolina to Argentina (See Relevant literature section). *Lutraria canaliculata* Say, 1822, is a junior synonym. The types were not found at the ANSP collection. However, this name was mentioned and illustrated in several publications (i.e. Conrad 1831; De Kay 1843; Adams and Adams 1853–1858; Reeve 1854a; Chenu 1862; Perry 1940). *Mactra campechensis* of Gray (1825) is a synonym. It was described based on specimens collected in Campeche, Mexico. Type material examination of *Raeta perspicua* (NMNZ 0321, holotype, registered erroneously from Bay of Islands, New Zealand) confirmed it as a junior synonym of Lamarck species. *Mactra papyracea* Gmelin, 1791 is a possible older name (sensu Lamarck, 1818 and d'Orbigny, 1846 in 1834–1847). It was originally described by Chemnitz (1782, vol. 6, pl. 23, Fig. 231, rejected by nomenclatorial purposes, ICZN 1944). The type material was not found, and the Chemnitz's illustration is rudimentary to confirm this change.

RELEVANT LITERATURE

Anatomy and Morphology: Signorelli and Pastorino, 2012a: 95–97.

Ecological and Environmental studies: De Castro and Dos Santos, 1989: 103.

Paleontology: Aguirre, 1989, 1991, 1992; Figueiras and Broggi, 1972–1973: 221; Maury, 1920: 137.

Taxonomy: Adams and Adams, 1853–1858: 386; Altena, 1971: 54; Barattini and Ureta, 1961: 170; Camp et al., 1998: 12; Carcelles, 1944: 284; Carpenter, 1863: 368; Castellanos, 1970: 238; Coan and Valentich-Scott, 2012: 559–560;

Conrad, 1831: 46, 1868: 41; d'Orbigny, 1846 (in 1834–1847): 527; Dall, 1894c: 28, 1895b: 212; De Kay, 1843: 232; Deshayes and Milne-Edwards, 1835: 93; Figueiras and Sicardi, 1969: 368; Gray, 1837: 375, 1853: 43; Hanley, 1842 (in 1842–1856): 27; Harry, 1969; Hutton, 1873: 65; Lamarck, 1818 in 1815–1822: 470; Lamy, 1913: 347; Mayer, 1867: 37; Mikkelsen and Bieler, 2007: 372; Perry, 1940: 83; Reeve, 1854a: sp 122; Rios, 1966: 33, 1970: 197, 1975: 237, 1985: 244, 1994: 267, 2009: 538; Signorelli and Pastorino, 2012a: 95–97; Skoglund, 2000; Turgeon et al., 2009: 730; Vanatta, 1903: 757; Zelaya, 2016: 253.

Raeta (*Raeta*) *undulata* (Gould, 1851) (Fig. 5.1i–l)

Shell ovate-trigonal, subequilateral, fragile, elongated posteriorly; length to 110 mm; umbo prosogyrate, broad but small; external surface with strong commarginal undulations regular mostly over the middle area, less evident near the ventral margin; minute radial and commarginal striae; anterior end broadly rounded; posterior end more acute, depressed, set off by low ridge; periostracum thin, yellowish, dehiscent; interior white with soft wrinkles that reveal the external ornamentation, mainly in juvenile specimens, translucent; hinge plate with only one anterior and one posterior lateral tooth in both valves, very short and close to the cardinals, the left valve with the inverted V-shaped cardinal teeth closely flanked by the accessory lamella; chondrophore very well ventrally developed; pallial sinus deep, about 60% shell length.

Type locality: La Paz, Baja California, Mexico.

Type material: MCZ 169453, holotype.

Geographic distribution: West coast of America from California to Caleta Chipana, Tarapacá, Chile (Coan and Valentich-Scott, 2012).

Stratigraphic distribution: Lower Miocene—Recent.

Synonyms: *Raeta gibbosa* Gabb, 1869: 30; *Labiosa* (*Raeta*) *gabbi* Pilsbry and Johnson, 1917: 202; *Labiosa* (*Raeta*) *gardnerae* Spieker, 1922: 168; *Labiosa* (*Raeta*) *hasletti* Anderson, 1929:177.

Remarks: Four fossil species have been mentioned in the literature as junior synonyms of *Raeta undulata*. They are *Raeta gibbosa* (types not at ANSP, Olsson, 1932) described from Lower Pleistocene beds exposed at Payta, Peru; *Labiosa* (*Raeta*) *gabbi* (ANSP 2675, syntype) a new name for *R. canaliculata* Gabb, 1873 (*non* Say, 1822); *Labiosa* (*Raeta*) *gardnerae* (types not found) described from Zorritos Formation, northern Peru; and *Labiosa* (*Raeta*) *hasletti* (CAS 4171, holotype) described from Tubera Formation, Colombia. The analysis of original illustrations, as well as, type material whenever possible, suggested that all nominal species must be considered as junior synonyms of *R. undulata*.

Relevant literature

Anatomy and Morphology: Harry 1969; Taylor et al., 1973: 265.
Ecological and Environmental studies: Bizjack et al., 2017.

Paleontology: Arnold, 1903: 177; Cantalamessa et al., 2005; Hodson and Hodson, 1931: 21; Olsson, 1932: 131; Pilsbry, 1931: 427; Pilsbry and Olsson, 1941: 73; Stump, 1975; Woodring, 1982: 653–654.

Taxonomy: Bernard, 1983: 41; Cauquoin, 1969: 1029–1030; Coan and Valentich-Scott, 1997: 17; Coan and Valentich-Scott, 2012: 560; Coan et al., 2000: 465; Conrad, 1868: 42; Dall, 1894d: 41, 1916a: 39; Gemmell et al., 1987:45–46; Grant and Gale, 1931: 407; Harry 1969; Hertlein and Strong, 1950: 235, 1955: 205; Keen, 1971: 207; Lamy, 1918: 355; Olsson and Petit, 1964: 533–534; Olsson, 1961: 332; Paredes et al., 2016: 141; Ramírez et al., 2003, 268; Stearns, 1894: 157; Strong, 1925; Tomicic, 1978.

Chapter 6
The Family Mesodesmatidae (Mollusca: Bivalvia)

Abstract The living species distributed in the American continent belonging to the family Mesodesmatidae Gray, 1840 are redescribed in this chapter. Two valid genera are recognized. The genus *Mesodesma* Deshayes, 1832 In Bruguière et al., 1789–1832 includes three living species. They are *M.* (*Mesodesma*) *donacium* (Lamarck, 1818), *M.* (*Ceronia*) *arctatum* (Conrad, 1831) and *M.* (*Ceronia*) *deauratum* (Turton, 1822). Finally, the genus *Amarilladesma* Huber, 2010, recently described, includes the Southwestern Atlantic *A. mactroides* (Reeve, 1854c).

Keywords Mesodesmatidae · *Mesodesma* · *Ceronia* · *Amarilladesma*

Family Mesodesmatidae Gray, 1840

[nom. correct. Dall, 1895b pro Mesodesmidae [sic] Gray, 1840, p. 154]

Shell subequilateral, compressed, donaciform or cuneiform, slightly or strongly anteriorly extended; external surface smooth or with commarginal sculpture; mactroid hinge plate with two cardinal and two lateral teeth; pallial sinus present, U-shaped or without it; pedal protractor and pedal elevator muscle scars, completely divided; siphon type A+ (according to Yonge, 1982b).

Subfamily Mesodesmatinae Gray, 1840

[nom. transl. et correct Dall, 1895b, p. 557 ex Mesodesmidae [sic] Gray, 1840, p. 154; = Paphiadae [sic] Gray, 1847, p. 186, spelling corrected to Paphiidae by Bouchet and Rocroi, 2010, p. 64, invalid, based on *Paphia* Lamarck, 1799, a junior homonym of *Paphia* Röding, 1798 [a venerid]; ?Myadesmidae B. L. Clark, 1922, p. 115, based on *Myadesma* B. L. Clark, 1922, spelling corrected to Myadesmatidae by Bouchet and Rocroi, 2010, p. 57; Donacillidae MacPherson and Gabriel, 1962, p. 359]

J. H. Signorelli, *The Superfamily Mactroidea (Mollusca:Bivalvia) in American Waters*, https://doi.org/10.1007/978-3-030-29097-9_6

Shell medium size, more or less compressed, sub-trigonal to elongate-ovate, anteriorly narrowly ovate, wedge-shaped, posteriorly slightly truncate; sculpture smooth or commarginal striated; umbos low, slightly posterior to midline; anterior and posterior, finely serrated lateral teeth present in left valve; one anterior and two posterior, finely serrated lateral teeth present in right valve; pallial sinus present.

Genus Mesodesma *Deshayes, 1832 in Bruguière et al., 1789–1832*

[= *Amphidesma* of authors (*non* Lamarck, 1818, 489)]

Shell donaciform, posterior end short; hinge arched, with lateral teeth striate; external ligament short, internal placed in a chondrophore with raised margins; pallial sinus broad, well-marked but short; posterior end obliquely truncate.

TYPE SPECIES: *Mactra donacia* Lamarck, 1818, by subsequent designation (Lamy, 1912a).

GEOGRAPHIC DISTRIBUTION: Eastern Pacific Ocean from Peru to Chile, western Atlantic from Greenland to Virginia, United States.

REMARKS: Type species designation was recently revealed by Huber (2010). It was not Anton (1838), due to he only listed the species in his personal collection without selecting anything. It was Lamy (1912a) who correctly designated the type species of *Mesodesma*. Keen in Moore (1969) and other authors synonymized *Mesodesma* with *Ceronia* due to morphological similarities. However, the biogeographic difference plus the larger size of *M. donacium* suggest a subgeneric separation. Three American species are included in *Mesodesma*, one *Mesodesma s. s.* and two in the north Atlantic *Ceronia*.

Subgenus Mesodesma s. s.

Mesodesma (*Mesodesma*) *donacium* (Lamarck, 1818) (Fig. 6.1a, b)

Shell large, triangular, thick, solid, length to 100 mm, strongly inequilateral, equivalve, posterior end extremely short, truncate; umbos not inflated, posterodorsal margin short, anterodorsal margin straight, long, anterior end rounded; external surface smooth, shiny, yellowish periostracum, sometimes greenish; with fine and irregular growth lines; external ligament short and weak; internally white, porcelaneous, hinge plate with a spoon-like chondrophore, ventrally projected, deep; left valve with one anterior and one posterior lateral tooth (LAII and LPII), the anterior longer; right valve with two anterior (LAI and LAIII) and two posterior (LPI and LPIII) lateral teeth; all lateral teeth dorsally striated; cardinal teeth small, the anterior cardinals longer; the pallial sinus rather wide, broad, moderately deep one-third

Fig. 6.1 (**a**, **b**) *Mesodesma* (*Mesodesma*) *donacium* (Lamarck, 1818), MNHN-IM-2000-23203, holotype; (**c–f**) *Mesodesma* (*Ceronia*) *arctatum* (Conrad, 1831), ANSP 51199, lectotype. Scale bar (**a**, **b**) 4 cm, (**c–f**) 1 cm

shell length, pallial line not over the edge on the anterior side; adductor muscle scar large, the anterior scar lenticular, the posterior subcircular.

TYPE LOCALITY: Unknown.

TYPE MATERIAL: MNHN-IM 2000-23203, holotype.

GEOGRAPHIC DISTRIBUTION: West coast of South America from Peru to southern Chile.

STRATIGRAPHIC DISTRIBUTION: Upper Pleistocene—Recent.

SYNONYMS: *Donacilla chilensis* d'Orbigny, 1845: 530; ?*Mesodesma cuneata* Potiez and Michaud, 1844: 206 (*non* Deshayes and Milne Edwards, 1835); *Mesodesma lanceolata* Reeve, 1854c: pl. 4, sp. 25.

REMARKS: Type material examination of *Donacilla chilensis* (NHMUK 1854.12.4.688, syntypes) confirmed it as a junior synonym of *M. donacium*. *Mesodesma lanceolata* (NHMUK 1968496, syntypes) was considered by Lamy (1914b) as a juvenile of *M. donacium* with an erroneous type locality. Huber (2010) placed *M. lanceolata* into an undescribed genus living in Indonesian waters. However, after revision of type material, he corrected himself and restored *Mesodesma lanceolata* as a junior synonym of *M. donacium* (Huber, 2015). Finally, the types of *Mesodesma cuneata* have not been found. They are probably housed into the cabinets of the Museum of Douai, however, the original illustrations given by Potiez and Michaud (pl. 62, Fig. 2) fit very well with Chilean *Mesodesma*.

RELEVANT LITERATURE

Anatomy and Morphology: Carré et al., 2005, 2006.

Ecological and Environmental studies: Carstensen et al., 2006; Jaramillo et al., 1994; Marins and Levy, 1999; Riascos et al., 2007, 2009.

Fisheries: Díaz et al., 2008; Gallardo-Escárate et al., 2015; Joo and Dupré, 2002; Ortiz and Stotz, 1996, 2003; Peredo et al., 1987; Pérez and Chávez, 2004; Segura Zamudio et al., 1998.

Parasitology: Bahamonde and López, 1962; López et al., 2014.

Taxonomy: Amar et al., 2008; Beu, 1971; Dall, 1909: 161; Deshayes and Milne-Edwards, 1835: 106; Hanley, 1856 (in 1842–1856): 38; Huber, 2010: 455; Hupé, 1854: 352; Lamy, 1912a: 246, Lamy, 1914a, b: 15–16; Martínez and Saez, 1869: 20; Paredes et al., 2016: 141; Ramírez et al., 2003: 268; Reeve, 1854c: sp. 11.

Subgenus **Ceronia** *Gray, 1853*

Shell ovate, elongated, compressed, posterior end short, rounded, moderately truncated; lateral teeth similar in size, elongated, pallial sinus short, rounded; anterior adductor muscle scar lenticular.

TYPE SPECIES: *Erycina denticulata* Gray, 1825 (= *Mactra deaurata* Turton, 1822), by monotypy.

GEOGRAPHIC DISTRIBUTION: Eastern North America.

REMARKS: Type species of *Ceronia* and *Mesodesma* are similar in shell and hinge plate morphology. Several authors synonymized both generic names. However, Huber (2010) considered *Ceronia* as a subgenus of *Mesodesma* due to slight differences like larger size and the remarkable gaped posterior end in *M. donacium*. A genetic revision will confirm the validity of this distinction. At the moment, the subgenus *Ceronia*, includes the type species *Erycina denticulata* Gray, 1825 (= *M.* (*Ceronia*) *deaurata* (Turton, 1822)) and *M.* (*Ceronia*) *arctatum* (Conrad, 1831), both living along the north west Atlantic Ocean.

Mesodesma (*Ceronia*) *arctatum* (Conrad, 1831) (Fig. 6.1c–f)

Shell robust, thick and acutely truncate posteriorly; external surface with strong but irregular concentric ornamentation covered by a brownish periostracum, absent in the umbonal area; anteroventral end elongated and rounded; hinge plate strong and broad, one anterior and one posterior lateral tooth in each valve; two cardinal teeth, small, flanked posteriorly by the internal ligament; left posterior cardinal tooth larger and inserted between right cardinal when the valves are closed; lateral teeth with transversal serrations acting as a locking device; internal ligament placed in a triangular chondrophore below the small umbos; adductor muscle scars large, the anterior somewhat larger in size; pallial sinus shallow; siphons separated and retractable into shell; usually not extended beyond the margin of the valves, inhalant siphon with two rings of tentacles, the exhalant with only one ring; mantle fold fused, only opened to form the inhalant and exhalant siphons; ventral fusion of inhalant siphon extended to the ventral margin; foot highly muscular; outer and inner demibranchs present, the outer smaller than the inner; food groove only over the distal edge of the inner demibranchs; labial palps narrow and strap-like; the outer palp attached to the mantle; oesophagus long; style sac separate from the mid-gut large and elongate and extends posteroventrally as a continuation of the stomach; stomach moderately large.

TYPE LOCALITY: Massachusetts, United States.

TYPE MATERIAL: ANSP 51199, lectotype designated by Davis (1964), length 25.2 mm, height 18.0 mm.

GEOGRAPHIC DISTRIBUTION: Greenland to Chesapeake Bay, United States.

STRATIGRAPHIC DISTRIBUTION: Pleistocene—Recent.

SYNONYMS: None.

REMARKS: *Mesodesma* (*Ceronia*) *arctatum* is a well-known species distributed along the northern Atlantic coast of North America. No synonyms were found in the literature.

RELEVANT LITERATURE

Anatomy and Morphology: Allen, 1975.

Ecological and Environmental studies: Merrill et al., 1978.

Fisheries: Brêthes, et al., 1986.

Taxonomy: Abbott, 1991; Brunel et al., 1998: 129; Davis, 1963, 1964; Gould, 1841: 57; Lamy, 1914a, b: 17–18; Morris, 1951: 88; Stimpson, 1860: 3; Yonge and Allen, 1985.

Mesodesma (*Ceronia*) *deauratum* (Turton, 1822) (Fig. 6.2a–d)

Shell thick, oval, equivalve, inequilateral; anterior end rounded; posterior end not sharply truncate as in *Mesodesma arctatum*; dorsal margin is relatively straight, whereas ventral slightly curved; umbos almost placed at the half of shell length; external surface with concentric and conspicuous growth lines; periostracum present

Fig. 6.2 (**a–d**) *Mesodesma* (*Ceronia*) *deauratum* (Turton, 1822), USNM 172665, lectotype; (**e–h**) *Amarilladesma mactroides* Reeve, 1854c, NHMUK 1968493, syntypes. Scale bar (**a–d**) 5 mm, (**e–h**) 2 cm

over the entire surface; internally white; hinge plate with two lateral teeth on the left valve, the anterior longer than the posterior; right valve with one anterior and one posterior lateral groove where fit the left lateral teeth; internal ligament placed in a chondrophore below the umbos; pallial sinus shallow and U-shaped; anterior adductor muscle lenticular and posterior subcircular.

TYPE LOCALITY: Exmouth, England, probably incorrectly recorded or locally introduced by the ballast of ships. Later, Davis (1965) designated Petite Matane, on the southern shore of Saint Lawrence estuary, Quebec province, Canada, as type locality.

Type material: USNM 172665, lectotype, Jeffrey's collection, dry, two valves. Length 27.9 mm, height 17.1 mm.

Geographic distribution: restricted to the Saint Lawrence estuary and the Gulf of the Saint Lawrence.

Stratigraphic distribution: Pleistocene—Recent.

Synonyms: *Erycina denticulata* Gray, 1825: 135; *Mesodesma jauresii* De Joannis, 1834: pl. 54; *Mesodesma anodontina* Römer, 1862: 135.

Remarks: Type materials of the mentioned synonyms were not found, and their confirmation as junior synonyms of *M.* (*C.*) *deauratum* must be checked. *Erycina denticulata* was illustrated by Wood (1828) and Griffith and Pidgeon (1833) and it was considered a synonym of *M.* (*C.*) *deauratum* by several authors (Lamy, 1914a, b; Davis, 1965; Petit and Coan, 2008). *Mesodesma jauresii* was described based on specimens collected at Gulf of the Saint Lawrence, and its original illustration is very similar to *M.* (*C.*) *deauratum* examined specimens. Finally, the original description of *Mesodesma anodontina* suggested morphological similarities with *M.* (*C.*) *deauratum*.

Relevant literature

Ecological and Environmental studies: Hutcheson and Stewart, 1994; Kostylev et al., 2001; Savenkoff et al., 2007.

Taxonomy: Abbott, 1974; Beu, 1971; Brunel et al., 1998: 129; Davis, 1963, 1964, 1965; Gould, 1841: 58, 1870: 81; Gray, 1825: 135, 1853: 44; Lamy, 1914a, b: 19; Petit and Coan, 2008: 231; Reeve, 1854c: sp. 14; Stimpson, 1860: 3; Wood, 1828: Fig. 9.

Genus Amarilladesma *Huber, 2010*

Shell medium sized, fragile, umbos prominent, elongate rounded pallial sinus, passing midline, U-shaped; hinge plate with cardinal teeth as in *Mesodesma*, lateral teeth almost smooth, only weakly striate; external ligament small, internal placed in a chondrophore ventrally projected; strongly gaping at both ends; with a dehiscent yellowish periostracum.

Type species: *Mesodesma mactroides* Reeve, 1854c by original designation.

Geographic distribution: southwestern Atlantic Ocean, from Brazil to Bahía San Blas.

Remarks: Marins and Levy (1999) found high genetic distance between *A. mactroides* (Brazil) and *M. donacium* (Chile) and suggested that both species are not congeneric. Based on these conclusions, Huber (2010) recently described a new genus to include the southwestern Atlantic species *Amarilladesma mactroides* (Reeve, 1854c). The morphological differences to *Mesodesma* mentioned by the author were the texture, pallial sinus and hinge morphology dentition.

Amarilladesma mactroides (Reeve, 1854c) (Fig. 6.2e–h)

Shell, thin, fragile, slightly inflated, oval, inequilateral, white; anterior margin acute, posterior margin subtruncate; umbos low, placed close to the posterior end, than anterior; external surface with delicate concentric lines covered by a pellucid to yellowish periostracum; internally white, with anterior muscle scar lenticular, posterior oval; pallial sinus deep, U-shaped; hinge plate with two lateral teeth and two cardinal teeth with an accessory lamina in the left valve; and two cardinal and three lateral teeth in the right valve; external ligament not prominent; internal well developed placed in a sub-triangular chondrophore; siphons separated and long, about three times shell length; external surface with two rows of longitudinal papillae; inhalant siphon with three categories of tentacles; exhalant siphon with a single row of small tentacles; with siphonal hearts; mantle fused with three gapes, the pedal gape and the siphonal gapes, also fused between siphons; foot axe-shaped; ctenidia with two demibranchs, the inner longer than the outer; with a supra-axial extension of the outer demibranchs; type C of Atkins (1937); labial palps triangular and long; with inner surfaces deeply plicate and outer smooth; mouth followed by a long and ciliated oesophagus; stomach type V of Purchon (1960); style sac and mid-gut separated.

TYPE LOCALITY: "Australia" registered erroneously.
TYPE MATERIAL: NHMUK 1968493, Syntypes, three specimens.
GEOGRAPHIC DISTRIBUTION: South western Atlantic Ocean from Rio de Janeiro, Brazil to Bahía San Blas, Argentina.
STRATIGRAPHIC DISTRIBUTION: Pleistocene—Recent.
SYNONYMS: *Donacilla solenoides* d'Orbigny, 1845 (*non* King, 1832, as *Erycina*); *Mesodesma arechavalettoi* Pilsbry, 1897.
REMARKS: *Amarilladesma mactroides* is a well-known species living in the south western Atlantic Ocean. It was widely mentioned in the literature (See Relevant literature section). *Erycina solenoides* King, 1832 (currently *Darina solenoides*) was recombined by d'Orbigny (1845 in 1834–1847) in *Donacilla.* The examination of d'Orbigny description clearly suggest that he misidentified a previously established specific name. The specimens that d'Orbigny found along the coast of San Blas are in fact *Amarilladesma mactroides*, undescribed at that moment. However, according to the Article 49 of the code, the name of d'Orbigny cannot be used as an available name, for the new taxon, even if he included it in a different genus. The epithet mactroides prevails. The type material of *Mesodesma arechavalettoi* (ANSP 70486, syntypes) described from specimens collected in Mar del Plata, coincides with *A. mactroides*.
RELEVANT LITERATURE

Anatomy and Morphology: Cabrera, 1959, 1960a, b, 1962; Castellanos, 1948; De Moreno et al., 1976a, b; Lepore et al., 2009; Narchi, 1981.

Ecological and Environmental studies: Bastida et al. 1991; Brazeiro and Defeo, 1999; Carcedo et al., 2015, 2017; Carvalho et al., 2015; Cledón and Nuñez, 2010; Defeo and Scarabino, 1990; Defeo et al., 1986, 1992; Fiori and Carcedo,

2015; Fiori and Morsán, 2004; Fiori and Defeo, 2006; Fiori et al., 2004a, b, 2019; Giacomin et al., 2014; Herrmann et al., 2009, 2011; Laitano et al., 2019; Lima et al., 2000; Manta et al., 2016; Olivier et al., 1971; Rapoport, 1960; Thompson and Sánchez de Bock, 2009; Vázquez et al., 2016.

Fisheries: Coscaron, 1959; Dadon, 1998, 1999a, b; Dadon and Castaños, 1995; Dadon and Chiappini, 2000; Dadon et al., 1997, 2001; Defeo, 1985, 1989; Fiori and Cazzaniga, 1999; Masello and Defeo, 1986; Olivier and Penchaszadeh, 1968.

Parasitology: Brusa et al., 2006; Cremonte and Figueras, 2004; Vázquez and Cremonte, 2017.

Taxonomy: Carcelles, 1939; Deshayes, 1855b: 336; d'Orbigny, 1845: 529; Huber, 2010: 762; Lamy, 1912a, b: 312, 1914a, b: 26–30; Marins and Levy, 1999; Reeve, 1854c: sp 15; Scarabino et al. 2016; Zelaya, 2016.

Appendix A: List of All Nominal Recent Species Belonging to the Superfamily Mactroidea Distributed in American Waters

Valid species (in the current combination)	Synonym	Examined type material
Harvella elegans (G.B. Sowerby I, 1825)		NHMUK 20190673, two syntypes
	Harvella pacifica Conrad, 1867	ANSP 51308, syntype
	Mactra estrellana Olsson, 1922	PRI 21265, holotype
	M. (Harvella) sanctiblasii Maury, 1925	PRI 2354, holotype
	Raeta maxima Li, 1930	AMNH 268093, lectotype; AMNH 268093a, paralectotype
	Harvella elegans tucilla Olsson, 1932	PRI 2252, holotype
Mactrellona alata (Spengler, 1802)		ZMUC-BIV, holotype, articulated specimen; ZMUC-BIV, paratype, one complete specimen
	Mactra laevigata Schumacher, 1817	ZMUC-BIV 1036, holotype
	Mactra carinata Lamarck, 1818	MNHN-IM-2000-7038, syntypes
	Mactrellona concentrica (Bory de Saint Vincent, 1827, in Bruguière et al. 1791–1827)	Types not found, based on the figure of the "Tableau of Encyclopedique Methodique..." (pl. 251, Fig. 2a, b, pl. 252, Fig. 2c) published in 1797 without a nomenclatorial act
Mactrellona clisia (Dall, 1915)		USNM 271481, holotype
Mactrellona exoleta (Gray, 1837)		NHMUK 196327, syntype, one complete specimen

J. H. Signorelli, *The Superfamily Mactroidea (Mollusca:Bivalvia) in American Waters*, https://doi.org/10.1007/978-3-030-29097-9

Valid species (in the current combination)	Synonym	Examined type material
	Lutraria ventricosa Gould, 1851 (invalid: junior homonym of *Lutraria ventricosa* Goldfuss, 1841)	MCZ 169451, holotype; MCZ 169452, paratype; USNM 15902, paratypes, four valves
Mactra subalata Mörch, 1861		ZMUC-BIV one syntype
Mactromeris catiliformis (Conrad, 1867)		Types not found
	Spisula catiliformis alcatrazensis Arnold	USNM MO 165291, holotype
	Spisula mercedensis Packard, 1916	UCMP 12067, holotype
Mactromeris hemphillii (Dall, 1894a)		USNM 15815, syntype, two single valves; CAS 66541, probably syntype, specimen from Hemphill collection, collected at type locality
	Spisula camaronis Dall, 1921b	USMN MO 333117, syntype
	Spisula hemphillii orcutti Manger, 1934	Type not found
	Spisula mossbeachensis Glen, 1959	UCMP 37643, holotype, UCMP 37644-37646, paratypes
Mactromeris polynyma (Stimpson, 1860)		Replacement name for *Mactra ovalis* Gould, 1841 (not Sowerby, 1817)
	Mactra similis Wood, 1828 (*non* Say, 1822: 309)	Type not found
	Mactra grandis Deshayes, 1832 (*non* Gmelin, 1791: 3259)	MNHN-IM probable holotype
	Mactra ovalis Gould, 1841 (*non* G.B. Sowerby I, 1817: 136)	Mentioned as being in Massachusetts State Cabinet N° 229 and Boston Society of Natural history (BSNH) N° 2324 but both specimens lost (Johnson, 1964)
	Mactra ponderosa Philippi, 1844, in Philippi, 1844–1850 (*non* Eichwald, 1830: 207)	SMF 316281, syntypes
	Mactra grayana Schrenk, 1867	Type not found
	Callista voyi Gabb, 1869	UCMP 12993, holotype
	Mactra loebbeckeana Weinkauff, 1881	Type not found

Valid species (in the current combination)	Synonym	Examined type material
	Spisula alaskana Dall, 1894d (nomen nudum), validated in Dall (1921a)	USNM 133946, holotype
	Spisula vladivostokensis Bartsch, 1929	USNM 369067, holotype
	Spisula korolevae Zhidkova, 1972 (in Zhidkova et al. 1972)	Type not found
Mulinia bicolor Gray, 1837		NHMUK 196323, holotype, one left valve
Mulinia byronensis Gray, 1837		NHMUK 196324, five syntypes
	Cumingia clerii Adams, 1850	NHMUK catalogue number unknown, 1 syntype
	Mactra byroni Weinkauff, 1884	No types, new proposed names for *M. byronensis*
	Mactra byroneana Weinkauff, 1884	No types, new proposed names for *M. byronensis*
	Mactra jousseaumei Mabille and Rochebrune, in Rochebrune and Mabille, 1889	MNHN-IM
	Mactra pencana Philippi, 1893	MNHNS 224, holotype
	Mactra lotensis Philippi, 1893	MNHNS 225, holotype
	Mactra magellanica Philippi, 1893	MNHNS 216, syntype
Mulinia cleryana (d'Orbigny, 1846)		NHMUK 1854.12.4.670, three syntypes
	Mactra guadelupensis Récluz 1852	MNHN-IM-2000-5017, two syntypes
	Mulinea portoricensis Shuttleworth, 1856	NMBE 501777
	Mactra carinulata Reeve, 1854a	NHMUK 1854.67.10.2.55
	Gnathodon cantrainei Sowerby in Reeve, 1873	Type not found
	Mulinia branneri Dall, 1901	USNM 107829, holotype

Valid species (in the current combination)	Synonym	Examined type material
Mulinia edulis (King, 1832)		NHMUK 1837.1.12.1168–9 and 196325, three syntypes
	Mactra cojimbensis Anton, 1837	Type not found
	Mulinia typica Gray, 1837	NHMUK 20190678, syntype
	Mactra rodatzi Dunker, 1850	Type not found
	Mactra marcida Gould, 1850	USNM 5892
	Mactra cuneola Gould, 1850	USNM 5891
	Mactra gabbi Tryon, 1870	ANSP 51392, syntypes
	Mulinia antarctica Paetel, 1883: 201, "ex Dunker 1850"	Type not found
	Mactra coquimbana Philippi, 1893	MNHNS 212, holotype
Mulinia exalbida Gray, 1837		NHMUK 1837.12.1.3688, holotype and one paratype
	Mactra cibaria Philippi, 1893	MNHNS 206, holotype
	Mactra calbucana Philippi, 1893	MNHNS 232, holotype
	Mactra jonasi Philippi, 1893	MNHNS 229, syntype
Mulinia lateralis (Say, 1822)	*Mactra rostrata* Philippi, 1849	MNHNS 213, holotype
Mulinia levicardo (E. A. Smith, 1881)		NHMUK 1879.10.15.67, syntype
Mulinia modesta (Carpenter, 1864)		USNM 3943, syntypes, four specimens
	Mactra coloradoensis Dall, 1894b	USNM 36570, syntypes
	Mulinia coloradoensis acuta Dall, 1894b	USNM 117506, syntypes
Mulinia pallida (Broderip and G.B. Sowerby I, 1829)		NHMUK 1858.5.12.49, holotype
	Mulinia donaciformis Gray, 1837	NHMUK 196326, syntype
	Mactra angulata Reeve, 1854a	NHMUK 196660, syntype

Valid species (in the current combination)	Synonym	Examined type material
	Mactra goniata Deshayes, 1855a	Type not found
	Mactra lacinata Carpenter, 1856	NHMUK 19621117, holotype
	Mactra (Mulinia) bistrigata Mörch, 1860	ZMUC-BIV-001549
	Mulinia bradleyi Dall, 1894b	USNM 101998, syntypes
	Corbula altirostris Li, 1930	AMNH 268092, holotype
	Mulinia camina Pilsbry and Olsson, 1941	ANSP 13700, holotype, ANSP 78997, paratype
Rangia (Rangia) cuneata (G.B. Sowerby I, 1832)		NHMUK 1970097, one syntype
	Rangia cyrenoides Des Moulins, 1832	MHNBx 2009.21540.0 for the whole lot (and label that is not specimen specific) and MHNBx 2009.21540.1 for syntype 1, 2009.21540.2 for syntype 2 and 2009.21540.3 for the biggest specimen without type status
	Gnathodon nasutus Dall, 1884	USNM 106988, holotype
Rangia (Rangianella) flexuosa (Conrad, 1840)		ANSP 52653, syntype, a single right valve
	Gnathodon rostratum Petit de la Saussaye 1853a	MNHN-IM-2000-34897, syntype
	Gnathodon flexuosa petitianus Dall, 1894e	USNM 57668A, holotype
Rangia (Rangianella) mendica (Gould, 1851)		Type not found, probably lost. Neotype herein designated USNM 103899, one complete specimen from Mazatlán, Mexico
	Gnathodon trigonum Petit de la Saussaye 1853a, b	MNHN-IM-2000-34896, syntype
	Gnathodon truncatum Petit de la Saussaye 1853a, b	*nomen nudum*, in figured caption
Rangia (Rangianella) pontchartrainensis (Morrison, 1965)		USNM 635643, holotype, and approximately a 1000 paratypes USNM 635644

Valid species (in the current combination)	Synonym	Examined type material
Simomactra dolabriformis (Conrad, 1867)		ANSP 51411, syntype, two valves
Simomactra falcata (Gould, 1850)		USNM 5893, holotype
	Anisodonta pellucida Dall, 1916b	USNM 208475, holotype
Simomactra planulata (Conrad, 1837)		Type not found
	Spisula strongi Burch, 1945	USNM 573603, holotype
Spisula (Spisula) fuegiensis E. A. Smith, 1905		NHMUK 1905.8.11.7-10, four syntypes
	Mactra epidermia Philippi, 1893 (*non* Reeve, 1854a)	MNHNS 219, holotype
Spisula (Spisula) guidoi Signorelli and Scarabino, 2010		Holotype: articulated valves MNHNM 15526; Paratypes MNHNM 15527 (three single valve, two articulated valves); MACN-IN 37507 (1v); MNHN 21121 (1v); FURG 30536 (1v); USNM 878150 (1v). All collected in the type locality. MNRJ 9261 (2 lv in alcohol), 23 km southeast of Chuy, Uruguay (33°52′00″S, 53°10′6″W; 22 m)
Spisula (Spisula) isabelleana d'Orbigny, 1846		NHMUK 1854.12.4.667/1, lectotype, NHMUK 1854.12.4.667/2, 3, two paralectotypes, designated by Aguirre (1994)
Spisula (Spisula) marplatensis Doello Jurado, 1949		MACN-In 10307, holotype, MACN-In 10307-1, six paratypes
Spisula (Spisula) petitii d'Orbigny, 1846		NHMUK 1854-12-4-665, holotype
	Mactra symmetrica Reeve, 1854a	NHMUK 20190674, holotype
	Mactra richmondi Dall, 1894c	USNM 124774, syntype
Spisula (Spisula) williamsi (Berry, 1960)		CAS 043985, holotype dry; SBMNH 34019, paratype 1 valve, dry (previous number SSB 9849), collected in 1937
	Spisula adamsi Olsson, 1961	ANSP 218917, holotype
	Mactra (Simomactra) hoffstetteri Cauquoin, 1969	MNHN-IM-2000-34899, syntypes

Valid species (in the current combination)	Synonym	Examined type material
Spisula (Hemimactra) solidissima (Dillwyn, 1817)		Type not found, name based on Chemnitz specimens (1788: 350, pl. 170, figure 1656, work rejected by ICZN, 1944, Opinion 184)
	Mactra procera Lightfoot, 1786 (Solander ms, *nomen nudum*)	
	Mactra gigantea Lamarck, 1818	Objective, name based on Chemnitz specimens (1788: 350, pl. 170, figure 1656, work rejected by ICZN, 1944, Opinion 184)
	Mactra similis Say, 1822	Type not found
	Spisula acadica Matthew, 1874	Type not found
	Spissula peninsulae M. Smith, 1937	Type in the personal collection of the author
Spisula (Hemimactra) raveneli (Conrad, 1832)		Type not found
	Spisula sayii Gray, 1837	Type not found
Standella (Standella) fragilis (Gmelin, 1791)		Type not found, name based on Chemnitz specimens (1788: 350, pl. 170, figure 1656, work rejected by ICZN, 1944, Opinion 184)
	Lutraria candida Lamarck, 1818	MNHN-IM-2000-7036, two syntypes
	Mactra braziliana Lamarck, 1818	MNHN-IM-2000-5015, one syntype
	Mactra oblonga Say, 1822	ANSP 51418, one syntype
	Mactra bilineata Reeve, 1854a	NHMUK 20190676/1, /2, two syntypes
	Mactra anserina Guppy, 1875	Types not found
	Mactra gigas Nowell-Usticke, 1969	Types not found
Standella (Standella) nasuta (Gould, 1851)		USNM 2296, Syntypes, two articulated specimens
	Mactra hiantina Deshayes, 1855a	Type not found
	Mactra deshayesii Conrad, 1868 (*non* Mayer, 1867)	No types. New name for *Mactra californica* Reeve (1854a, sp. 114, *non* Conrad, 1837)

Valid species (in the current combination)	Synonym	Examined type material
	Mactrotoma revellei Durham, 1950	UCMP 32237, holotype
Standella (Micromactra) angusta (Reeve, 1854a)		NHMUK 20190677, holotype
	Mactra (Micromactra) atacama Pilsbry and Olsson, 1941	ANSP 175548, paratype
Standella (Micromactra) californica (Conrad, 1837)		Type not found
Standella (Micromactra) janeiroensis (E. A. Smith, 1915)		NHMUK 1915.4.18.489, six Syntypes
Standella (Micromactra) surinamensis (Altena, 1968)		RMNH.MOL.55788, holotype, a left valve (Fig. 151a–c) from Surinam (no exact locality, but the specimen is obviously Recent), June 1910, collected by D. G. J. Bolten. Seven paratypes mentioned in the original description. Holocene paratypes (three lots): RMNH.MOL.55794, one right valve from Holocene deposits from Coronieweg, picket, Coronie District, collected by D. C. Geijskes; RMNH.MOL.55793, one left valve (Fig. 151d) from Coronieweg, km 117–118, Coronie District, collected by C.O v. R. Altena; RMNH.MOL.55798, one small right valve from Charlesburg, Surinam district, collected by D. C. Geijskes. Recent Paratypes (four lots): RMNH.MOL.55795, one left valve from Suriname, collected by D. C. Geijskes; RMNH.MOL.55792, a right and a left valve (Fig. 152a, b) from Coppename Punt, Saramacca district, collected by D. C. Geijskes; RMNH.MOL unnumbered, a small right valve from the beach near Popogaimama Creek, Saramacca district, collected by D. C. Geijskes; USNM 638959, one young right valve from Coquette station 2, 6°23′N, 55° 05.5′W, 15 fathoms, 5/11/1957 (mentioned in the original description but not found)
	Micromactra miskito Petuch, 1998	CMNH 47364, holotype; CMNH 47352, paratype

Valid species (in the current combination)	Synonym	Examined type material
Standella (Mactroderma) isthmica (Pilsbry and Lowe, 1932)		ANSP 130402, Holotype; ANSP 398872 (ex 130402), one paratype
	Mactra (Micromactra) vanattae Pilsbry and Lowe, 1932	ANSP 155921, syntype
	Mactra (Micromactra) vanattae acymata Pilsbry and Lowe, 1932	ANSP 152560, two syntypes
	Mactra (Micromactra) fonsecana Hertlein and Strong, 1950	CAS 61534, holotype; CAS 61535, 64511 and 159007, paratypes
Standella (Mactroderma) velata (Philippi, 1849)		MNHNS 101194, holotype, a single right valve
	Mactra paitensis Philippi, 1893	MNHNS 228, syntype
Trinitasia iheringi (Dall, 1897)		USNM 107632, holotype
	Mactra kempfi Cauquoin, 1969	MNHN-IM-2000-34898, syntype
	Mactra inceri Petuch, 1998	CMNH 47338, holotype, single right valve; CMNH 47346, 47350, paratypes
Tumbeziconcha goniocyma (Pilsbry and Lowe, 1932)		ANSP 155567, holotype
Tunmbeziconcha thracioides (Adams and Reeve, 1850)		NHMUK 1874.12.11.393, holotype
Tresus allomyax Coan and Valentich Scott, 2000 (in Coan et al. 2000)		CAS 093991, holotype, articulate specimen. CAS 099254, paratype from type localities. CAS 093994 and 093995, additional paratypes from off Clam beach, and from off Orick, respectively, both in Humboldt county, California
Tresus capax (Gould, 1850)		USNM 3884, holotype
	Lutraria maxima Middendorff, 1849 (invalid: junior homonym of *Lutraria maxima* Jonas, 1844)	ZISP 213/1, collected at Sitcha, ZISP 214/2 collected at Columbia River, syntypes

Valid species (in the current combination)	Synonym	Examined type material
Tresus nuttallii (Conrad, 1837)		NHMUK 1861.5.20.134, holotype
	Lutraria inflata Dunker, 1853	Type not examined
	Spisula longa Dall, 1921b	USNM 333116, syntypes
	Schizothaerus nuttallii bighopensis Henderson, 1931	Type not found
Darina solenoides (King, 1832)		NHMUK 1837.12.1.879/884, 1968507 and 1859.9.19.59, nine syntypes
	Lutraria kingi Fischer, 1887	No types, unnecessary new name for *D. solenoides*
	Lutraria tenuis Philippi, 1845	ZMB 112740, syntypes
	Darina declivis Carpenter, 1864	Redpath Museum 101, holotype
Darina rustica Huber, 2010		NHMUK 20100578, holotype; Paratypes 1 (30.1 mm, RA115) and 2 (32.2 mm, RA115 1) in collection M. Huber; Paratypes 3 (42 mm) and 4 (33 mm) in collection A. F. Suarez; Paratypes 5 (35 mm) and 6 (34 mm) in collection T. Narosky
Anatina anatina (Spengler, 1802)		ZMUC-BIV, two syntypes
	Anatina pellucida Schumacher, 1817	Unnecessary substitute name for *Mactra anatina*
	Lutraria lineata Say, 1822	ANSP 51313, syntypes
	Mactra papyracea Lamarck, 1818 (*non* Gmelin, 1791)	Type not found
	Mactra recurva Wood, 1828	NHMUK 20100621, holotype
	Mactra nuttallii Reeve, 1854a (vol. 8, pl. 21, sp. 125, *non* Conrad, 1837)	Type not found
Anatina cyprinus (Wood, 1828)		NHMUK 2010.06.08, holotype
Raeta plicatella (Lamarck, 1818)		MNHN-IM 2000-5011, two syntypes
	Lutraria canaliculata Say, 1822	Type not found
	Mactra campechensis Gray, 1825	Type not found
	Raeta perspicua Hutton, 1873	NMNZ 0321, holotype

Valid species (in the current combination)	Synonym	Examined type material
Raeta undulata (Gould, 1851)		MCZ 169453, holotype
	Raeta gibbosa Gabb, 1869	Type not found
	Labiosa (Raeta) gabbi Pilsbry and Johnson, 1917	ANSP 2675, syntype
	Labiosa (Raeta) gardnerae Spieker, 1922	Type not found
	Labiosa (Raeta) hasletti Anderson, 1929	CAS 4171, holotype
Mesodesma (Mesodesma) donacium (Lamarck, 1818)		MNHN-IM 2000-23203, holotype
	Donacilla chilensis d'Orbigny, 1845	NHMUK 1854.12.4.688, syntypes
	Mesodesma cuneata Potiez and Michaud, 1844	Not found, probably in the cabinets of the Museum of Douai
	Mesodesma lanceolata Reeve, 1854c	NHMUK 1968496, syntypes
Mesodesma (Ceronia) arctatum (Conrad, 1831)		ANSP 51199, lectotype
Mesodesma (Ceronia) deauratum (Turton, 1822)		USNM 172665, lectotype
	Erycina denticulata Gray, 1825	Type not found
	Mesodesma jauresii De Joannis, 1834	Type not found
	Mesodesma anodontina Römer, 1862	Type not found
Amarilladesma mactroides (Reeve, 1854c)		NHMUK 1968493, Syntypes
	Donacilla solenoides d'Orbigny, 1845 (*non* King, 1832)	No Types, misidentification of *Darina solenoides*
	Mesodesma arechavalettoi Pilsbry, 1897	ANSP 70486, syntypes

References

Abarca A, Oliva D, Gutiérrez R, Celis Á, Durán LR (2012) Grown-out of seeds of the taquilla clam *Mulinia edulis* (King and Broderip, 1832) in the subtidal zone in northern Chile and in the intertidal zone in southern Chile. Lat Am J Aquat Res 40(3):694–704

Abarca A, Toledo P, Oliva D (2018) Composición química de ejemplares de la almeja taquilla *Mulinia edulis* de bancos naturales del norte y sur de Chile. Rev Biol Mar Oceanogr 53(2):265–272

Abbott RT (1954) American Seashells. New york: D. Van Nostrand Company, Inc

Abbott RT (1974) American seashells. The marine Mollusca of the Atlantic and Pacific coasts of North America, 2nd edn. Van Nostrand Reinhold Ltd., New York

Abbott RT (1991) Seashells of the northern hemisphere. Gallery Books, New York

Abbott RT, Dance SP (1986) Compendium of seashells, 3rd Printing edn. American Malacologists Inc., Melbourne, FL

Abbott DP, Hilgard GH (1987) Observing marine invertebrates. Stanford University Press, Stanford, CA

Abraham A, Plakas SM, Flewelling LJ, El Said KR, Jester EL, Granadem HR et al (2008) Biomarkers of neurotoxic shellfish poisoning. Toxicon 52(2):237–245

Adams A (1850) Description of new species of the genus *Cumingia*, with some additional generic characters. Proc Zool Soc Lond 18(202):24–25

Adams H, Adams A (1853–1858) The genera of recent mollusca; arranged according to their organization. John van Voorst, Paternoster Row, London

Adams A, Reeve, LA (1850) Mollusca. p. i-x + 45-87. In. Adams A. (Editor) 1848-1850. The zoology of the voyage of H. M.S. Samarang, under the command of Captain Sir Edward Belcher, during the years 1843-1846. London. x + 87 pp.

Addicott WO (1963) Interpretation of the invertebrate fauna from the Upper Pleistocene Battery Formation near Crescent City, California. Proc Calif Acad Sci 31(13):341–347

Agassiz L (1860). Contributions to the Natural History of the United States of America 3:1–301

Aguirre M (1989) Holocene macrobenthic molluscan associations from northeastern Buenos Aires Province, Argentina. Quat South Am Antarctic Peninsula 7:161–195

Aguirre M (1991) Asociaciones de Moluscos Bentónicos Marinos del cuaternario tardío en el Noroeste Bonaerense. Ameghiniana 27:161–177

Aguirre M (1992) Caracterización faunística del Cuaternario marino del noreste de la Provincia de Buenos Aires. Rev Asoc Geol Argent 47:31–54

Aguirre M (1994) Revisión de las especies de *Mactra* Linné (Bivalvia) del Cuaternario Bonaerense (Argentina). Aspectos ecológicos y de distribución. Rev Esp Paleontol 9:131–144

Aguirre M (2003) Late Pleistocene and Holocene palaeoenvironments in Golfo San Jorge, Patagonia: molluscan evidence. Mar Geol 194(1–2):3–30

J. H. Signorelli, *The Superfamily Mactroidea (Mollusca:Bivalvia) in American Waters*, https://doi.org/10.1007/978-3-030-29097-9

Aguirre M, Savalscky JCH, Dellatorre F (2008) Late Cenozoic invertebrate paleontology of Patagonia and Tierra del Fuego, with emphasis on molluscs. Dev Quat Sci 11:285–325

Alamo V, Valdivieso V (1997) Lista sistemática de moluscos marinos del Perú, 2nd edn, revisada y actualizada edn. Boletín del Instituto Marino, Perú-Callao. 183 pp

Aldea C, Valdovinos C (2005) Moluscos del Intermareal roco del centro-sur de Chile (36°–38°S): taxonomía y clave de identificación. Gayana 69:364–396

Allen RD (1953) Fertilization and artificial activation in the egg of the surf-clam, *Spisula solidissima*. Biol Bull 105(2):213–239

Allen JA (1975) The functional morphology of *Mesodesma arcatum* (Conrad) (Bivalvia: Mesodesmatidae). Proc Malacol Soc Lond 41:601–609

Allen JA (1985) The recent Bivalvia: their form and evolution. In: Clarke ERTMR (ed) The Mollusca, Evolution, vol 10. Academic, London, pp 337–403

Altena COvR (1968) The Holocene and recent marine bivalve mollusca of Surinam. Stud Fauna Suriname Other Guyanas 10:152–178

Altena COvR (1969) The marine Mollusca of Suriname (Duch Guiana) Holocene and Recent. Part 1. Zool Verh 101:1–49

Altena COvR (1971) On six species of marine mollusca of Suriname, four of which are new. Zool Verh 119:51–54

Amar G, Palma Rojas C, Von Brand E, Jara-Seguel P (2008) Karyotype study in the surf clam *Mesodesma donacium* Lamarck, 1818 (Bivalvia: Veneroida: Mesodesmatidae). Gayana 72(1):18–22

Amaratunga T, Rowell TW (1988) Age and meat yield of Stimpson's surf clam, *Spisula polynyma*, a recently found commercial bivalve resource in eastern Canada. J Shellfish Res 7(1):107–108

Anderson FM (1929) Marine Miocene and related deposits of North Colombia. Proc Calif Acad Sci 18(4):73–213

Andrews J (1971) Shells and shores of Texas. University of Texas Press, Austin

Andrus C, Fred T, Rich KW (2008) A preliminary assessment of oxygen isotope fractionation and growth increment periodicity in the estuarine clam *Rangia cuneata*. Geo-Mar Lett 28(5–6):301–308

Anton HE (1837) Diagnosen einiger neuen Conchylien-Arten. Arch für Nat 3(1):281–286

Anton HE (1838) Verzeichniss der Conchylien, welche sich in der Sammlung von Hermann Eduard Anton befinden. Herausgegeben von dem Besitzer, Halle

Armstrong DA, Armstrong JL (1974) A haplosporidan infection in gaper clams, *Tresus capax* (Gould), from Yaquina Bay, Oregon. Proc Nat Shellfish Assoc 64:68–72

Arnold R (1903) The paleontology and stratigraphy of the marine Pliocene and Pleistocene of San Pedro, California. Mem Calif Acad Sci 3:1–420

Arnold R (1907) New and characteristic species of fossil mollusks from the oil-bearing Tertiary formations of Santa Barbara County, California. Smithson Misc Collect 50:419–447

Atkins D (1937) On the ciliary mechanisms and interrelationships of lamellibranchs Part III: Types of lamellibranch gills and their food currents. Q J Microsc Sci 2:375–421

Avellanal MH, Jaramillo E, Clasing E, Quijón P, Contreras H (2002) Reproductive cycle of the bivalves *Ensis macha* (Molina, 1782) (Solenidae), *Tagelus dombeii* (Lamarck, 1818) (Solecurtidae), and *Mulinia edulis* (King, 1831) (Mactridae) in southern Chile. Veliger 45(1):33–44

Avila-Serrano GE, Flessa KW, Téllez-Duarte MA, Cintra-Buenrostro CE (2006) Distribución de la macrofauna intermareal del Delta del Río Colorado, norte del Golfo de California, México. Cienc Mar 32(4):649–661

Bahamonde N, López MT (1962) *Proboscidosaccus mesodesmatis* n. sp. parásito de *Mesodesma donacium* Lamarck. Invest Zool Chilenas 8:43–56

Balech E, Ehrlich MD (2008) Esquema Biogeográfico del Mar Argentino. Rev Invest Desarrollo Pesquero 19:45–75

Barattini LP, Ureta EH (1960) La fauna de las costas uruguayas del este (Invertebrados). Museo Damaso Antonio Larranaga, Montevideo

Barattini, LP, Ureta EH (1961a) La fauna de las costas uruguayas del este (Invertebrados). Museo Damaso Antonio Larranaga, Montevideo

Barnard KH (1963–1974) Contributions to the knowledge of South African marine Mollusca. Part III. Gastropoda: Prosobranchiata: Taenioglossa (published February 1963). Ann South African Mus 47:1–201

Barnes PAG, Morton B (1997) The functional morphology of *Mactrinula reevesii* (Bivalvia: Mactroidea) in Hong Kong: adaptations for a deposit-feeding lifestyle. J Zool 241(1):13–34

Bartsch P (1915) Report on the Turton Collection of South African marine mollusks: with additional notes on other South African shells contained in the United States National Museum. US Government Printing Office, Washington, DC

Bartsch P (1929) Report upon the collection of marine mollusks made by Professor Dr. K. Derjugin in the Gulf of Peter the Great. Leningrad, Gosudarstvennyi Gidrologicheskii Institut, Issledovaniia Morei SSSR 10:129–140

Bastida R, Roux A, Bremec C, Gerpe M, Defeo O (1991) Estructura poblacional de la almeja amarilla *Mesodesma mactroides* durante el verano de 1989 en la provincia de Buenos Aires, Argentina. Frente Marit 9:83–92

Bedford WB, Anderson JW (1972) Adaptive mechanisms of the estuarine bivalve *Rangia cuneata* to a salinity stressed environment. Am Zool 12:721

Belding DL (1910) The growth and habits of the sea clam (*Mactra solidissima*). Report for Massachusetts, Commissioners on Fisheries and Game 1909, pp 26–41

Bell JL, Grassle JP (1998) A DNA probe for identification of larvae of the commercial surfclam (Spisula solidissima). Mol Mar Biol Biotechnol 7(2):127–137

Benson WH (1858) Characters of *Tanysiphon*, a New Genus of Fluviatile Shells, Allied to the Myacidae. Annals and Magazine of Natural History, series 3 2:407–410

Bernard FR (1983) Catalogue of the living Bivalvia of the eastern Pacific Ocean: Bering Strait to Cape Horn. Can Spec Publ Fish Aquat Sci 61:1–102

Bernard FR, Cai YY, Morton B (1993) A Catalogue of the Living Marine Bivalve Molluscs of China (Vol. 1). Hong Kong University Press

Berry SS (1960) Notes on new eastern Pacific Mollusca—IV. Leaflets Malacol 1(19):115–122

Berú P, Perea A, Gálvez MA (2014) First record of *Mactotoma velata* (Bivalvia, Mactridae) in Ancash—Peru, biological and reproductive aspects. Biologist (Lima) 12(2):207–221

Beu A (1966) The molluscan genera Lutraria, *Resania* and *Zenatia* in New Zealand. Trans R Soc N Z Zool 8:63–91

Beu A (1971) Genera of the bivalve family Mesodesmatidae, with comments of some Australasian species. J Malacol Soc Aust 2:113–131

Beu A (2006) Marina Mollusca of oxygen isotope stages of the last two million years in New Zealand. Part 2: Biostratigraphically useful and new Pliocene to recent bivalves. J R Soc N Z 36:151–338

Beu A, Rooij-Schuiling LA (1982) Subgeneric classification of New Zealand and Australian species of Paphies Lesson (Bivalvia: Mesodesmatidae), and names for the two species of tuatua in New Zealand. N Z J Zool 9:211–229

Bieler R, Carter JG, Coan E (2010) Classification of bivalve families. In P. Bouchet and J.-P. Rocroi (editors), 2010. Nomenclator of bivalve families. Malacologia 52:1–184

Bittner A (1892) Ueber die systematische Stellung von *Mactra* und verwandten Gattungen. Verhandlungen for Kaiserlich-Königlichen Geologischen Reichsanstalt 1892(9):232–241

Bizjack MT, Kidwell SM, Velarde RG, Leonard-Pingel J, Tomašových A (2017) Detecting, sourcing, and age-dating dredged sediments on the open shelf, southern California, using dead mollusk shells. Mar Pollut Bull 114(1):448–465

Blainville HDd (1824) Mollusques. In F. Cuvier & "plusieurs Professeurs du Jardin du Roi, et des principales Écoles de Paris", 1816-1830, Dictionnaire des Sciences Naturelles dans lequel on Traite Méthodiquement des Différens Êtres de la Natur, Considérés soit en Eux-mêmes, d'après l'état Actuel de nos Connoissances, soit Relativement a l'Utilité qu'en Peuvent Retirer la Médecine, l'Agriculture, le Commerce et les Arts. Suivi d'une Biographie des Plus Célèbres Naturalistes. F. G. Levrault, Strasbourg and Paris. vol. 32:1–392

Breed-Willecke GM, Hancock DR (1980) Growth and reproduction of subtidal and intertidal populations of the gaper clam Tresus capax (Gould) from Yaquina Bay, Oregon. National Shellfisheries Association, Proceedings 70(1):1–13
Bruguière JG, Lamarck JBPAdM, Bory de Saint Vincent JBGM (1791-1827) Tableau encyclopédie et méthodique de trois règnes de la nature. Vers, coquilles, mollusques et polypiers, Paris (Agasse). 3 volumes
Bruguière JG, Deshayes GP, Hwass CH (1789–1832) Encyclopédique Méthodique. Histoire Naturelle des Vers. 3 vols. Vol. 1, Panckoucke, Paris; vols. 2 and 3, H. Agasse, Paris
Bosc LAG (1816) Anatine. In: Nouveau dictionarire d'histoire naturelle. Deterville, Paris
Bosch DT, Dance SP, Moolenbeek RG, Oliver PG (1995) Seashells of Eastern Arabia. Motivate Publishing, Dubai
Boschi EE (1979) Geographic distribution of Argentinian marine decapod crustacean. Bull Biol Soc Wash 3:134–143
Boschi EE (2000a) Species of decapod crustaceans and their distribution in the American marine zoogeographic provinces. Rev Invest Desarrollo Pesquero 13:7–136
Boschi EE (2000b) Biodiversity of marine decapod Brachyurans of the Americas. J Crustac Biol 20:337–342
Boshoff PH, Field JG (1968) A report on some Pelecypoda dredged off the Natal and Moçambique coasts. Trans R Soc S Afr 38:79–94
Boss KJ (1982) Mollusca. In: Parker SP (ed) Synopsis and classification of living organisims. McGraw-Hill Book Company, New York
Bouchet P, Nicklès M (1976) *Mactra diolensis*, bivalve nouveau du Parc National de basse Casamance. Bull Inst Fondamental D'Afrique Noire 38A(1):57–61
Bouchet P, Rocroi JP (2010) Nomenclator of bivalve families; with a classification of bivalve families by R. Bieler, J. G. Carter, and E. V. Coan. Malacologia 52(2):1–184
Bourne N, Smith DW (1972a) The effect of temperature on the larval development of the horse clam, *Tresus capax* (Gould). Proc Nat Shellfish Assoc 62:35–38
Bourne N, Smith DW (1972b) Breeding and growth of the horse clam, *Tresus capax* (Gould), in southern British Columbia. Proc Nat Shellfish Assoc 62:38–46
Bourne N, Smith DW (1972c) Breeding and growth of the horse clam, *Tresus capax* (Gould), in southern British Columbia, Canada. Fisheries Research Board of Canada, Technical Report, vol 232, pp 1–42
Bousfield EL (1964) Coquillages des côtes canadiennes de l'Atlantique. National Museum of Canada, Ottawa
Branch GM, Griffiths CL, Branch ML, Beckley L (2010) Two oceans: a guide to the marine life of southern Africa. Struik Nature, Cape Town
Brazeiro A, Defeo O (1999) Effects of harvesting and density dependence on the demography of sandy beach populations: the yellow clam *Mesodesma mactroides* of Uruguay. Mar Ecol Prog Ser 182:127–135
Breed-Willecke G, Hanckok M, Hancock DR (1980) Growth and reproduction of subtidal and intertidal populations of the gaper clam *Tresus capax* (Gould) from Yaquina Bay, Oregon. Proc Natl Shellfisheries Assoc 70(1):1–13
Brêthes JCF, Desrosiers G, Fortin G Jr (1986) Croissance et production du bivalve *Mesodesma arctatum* (Conrad) sur la côte Nord du golfe du Saint–Laurent. Can J Zool 64(9):1914–1919
Bright J, Cohen AS, Starratt SW (2018) Distinguishing brackish lacustrine from brackish marine deposits in the stratigraphic record: a case study from the late Miocene and early Pliocene Bouse Formation, Arizona and California, USA. Earth-Sci Rev 185:974–1003
Broderip WJ, Sowerby IGB (1829) Observations on new or interesting Mollusca contained, for the most part, in the Museum of the Zoological Society. Zool J 4:359–379
Brown AC (2002) Extension to the known range of *Scissodesma spengleri* (Linné, 1767) (Bivalvia: Mactridae): news and views. S Afr J Sci 98:419
Brown LA, Furlong JN, Brown KM, Peyre MK (2013) Oyster reef restoration in the northern Gulf of Mexico: effect of substrate and age on ecosystem services. Restor Ecol 22(2):127–269

Bruguière JG, Deshayes GP, Hwass CH (1789–1832) Encyclopédique Méthodique. Histoire Naturelle des Vers. 3 vols. Vol. 1, Panckoucke, Paris; vols.2 and 3, H. Agasse, Paris

Bruguière JG, de Monet de Lamarck JBPA, de Saint Vincent JBGMB (1791–1827) Tableau encyclopédie et méthodique de trois règnes de la nature. Vers, coquilles, mollusques et polypiers. Agasse, Paris. 3 volumes

Brunel P, Bossé L, Lamarche G (1998) Catalogue of the marine invertebrates of the Estuary and Gulf of Saint Lawrence. Can Spec Publ Fish Aquat Sci 126:1–405

Brusa F, de León RP, Damborenea C (2006) A new *Paravortex* (Platyhelminthes, Dalyellioida) endoparasite of *Mesodesma mactroides* (Bivalvia, Mesodesmatidae) from Uruguay. Parasitol Res 99(5):566–571

Brusca RC, Thomson DA (1975) Pulmo Reef: the only coral reef in the Gulf of California. Cienc Mar 2(2):37–53

Brusca RC, Álvarez-Borrego S, Hastings PA, Findley LT (2017) Colorado River flow and biological productivity in the Northern Gulf of California, Mexico. Earth-Sci Rev 164:1–30

Burch T (1945) Additional notes on West American *Spisula* with a description of a new species. Minutes 49:50–51

Cable WD (1973) The vascular membrane in young Mactrid clams, *Spisula solidissima*. Nautilus 87:110–111

Cabrera SE (1959) Crecimiento de la almeja amarilla *Mesodesma Mactroides* (Desh) de la Costa bonaerense (Partido de Gral. Lavalle). Primer Congr Sudamer Zool 2:79–84

Cabrera SE (1960a) Estado Juvenil de la almeja amarilla Mesodesmatidae: *Mesodesma mactroides* Desh. Primer Congr Sudamer Zool 2:233–241

Cabrera SE (1960b) Regeneración de las valvas de almeja amarilla (*Mesodesma mactroides* Desh). Primer Congr Sudamer Zool 2:87–90

Cabrera SE (1962) Contribución al conocimiento del crecimiento de la almeja amarilla (*Mesodesma Mactroides* Desh) de la Costa bonaerense. Secretaría de estado de agricultura y ganadería de la nación, Dirección General de Pesca, La Plata, pp 1–15

Cain TD (1975) Reproduction and recruitment of the brackish water clam *Rangia cuneata* in the James River, Virginia. Fish Bull 73(2):412–430

Camp DK, Lyons WG, Perkins TH (1998) Checklists of selected shallow-water marine invertebrates of Florida. Florida Marine Research Institute Technical Report TR-3

Campbell A, Bourne N (2001) Population biology of gaper (horse) clams, *Tresus capax* and *T. nuttallii*, in southern British Columbia, Canada. J Shellfish Res 19(2):933–942

Campbell A, Bourne N, Carolsfeld W (1990) Growth and size at maturity of the Pacific gaper *Tresus nuttallii* (Conrad 1837) in southern British Columbia. J Shellfish Res 9(2):273–278

Campbell BN, Groot JB, Mahannah SM (2009) An investigation into ageing methods for horse clams (*Tresus nuttallii* and *T. capax*). Can Tech Rep Fish Aquat Sci 2765:1–25

Cantalamessa G, Di Celma C, Ragaini L (2005) Sequence stratigraphy of the Punta Ballena Member of the Jama Formation (Early Pleistocene, Ecuador): insights from integrated sedimentologic, taphonomic and paleoecologic analysis of molluscan shell concentrations. Palaeogeogr Palaeoclimatol Palaeoecol 216(1–2):1–25

Cantera J, Contreras R (1988) Bivalvos perforadores de esqueletos de corales escleractiniarios en la Isla de Gorgona, Pacífico Colombiano. Rev Biol Trop 36:151–158

Carcedo MC, Fiori SM, Piccolo MC, Abbate MCL, Bremec C (2015) Variations in macrobenthic community structure in relation to changing environmental conditions in sandy beaches of Argentina. Estuar Coast Shelf Sci 166:56–64

Carcedo MC, Fiori SM, Bremec C (2017) Zonation of macrobenthos across a mesotidal sandy beach: variability based on physical factors. J Sea Res 121:1–10

Carcelles A (1939) *Pectunculus longior* y *Mesodesma mactroides* de la Argentina y Uruguay. Physis 17:736–743

Carcelles A (1944) Catálogo de los moluscos marinos de Puerto Quequén. Rev Museo La Plata Zool 3:233–309

Carcelles A (1950) Catálogo de los moluscos marinos de Patagonia. Anales del Museo Nahuel Huapi 8:41–100. 2 extra nueva serie

Carcelles A, Williamson S (1951) Catálogo de los moluscos marinos de la provincia magallánica. Revista del Instituto Nacional de Investigación de las Ciencias Naturales Ciencias Zoológicas 2:225–383

Cargnelli LM, Griesbach SJ, Packer DB, Weissberger E (1999) Atlantic Surfclam, *Spisula solidissima*, life history and habitat characteristics. NOAA Technical Memorandum NMFS-NE-142

Carpenter PP (1855–1857) Catalogue of the reigen collection of the Mazatlan mollusca in the British Museum. Oberlin Press, Warrington. xvi + 552 p. [pp 1–120 published 1855; pp 121–444 published 1856; pp 445–552 published 1857]

Carpenter PP (1860) Check list of the shells of North America: No. 2(b). West coast: Mexican and Panamic Province. In I. Lea, P.P. Carpenter, W. Stimpson, W.G. Binney, and Temple Prime, Check lists of the shells of North America. Smithsonian Miscellaneous Collections 2(128)[(6)(2)]:1–13

Carpenter PP (1863) Review of Prof. C. B. Adams' catalogue of the shells of Panama, from the type specimens. Proc Zool Soc Lond 41:339–369

Carpenter PP (1864) Supplementary report on the present state of our knowledge with regard to the mollusca of the west coast of North America. Rep Br Assoc Advance Sci 1863:517–686

Carpenter KE (2002) The living marine resources of the western Central Atlantic. Volume 1: Introduction, molluscs, crustaceans, hagfishes, sharks, batoid fishes, and chimaeras. In: FAO species identification guide for fishery purposes and American Society of Ichthyologists and Herpetologists Special Publication, vol 5. Rome, FAO

Carré M, Bentaleb I, Blamart D, Ogle N, Cardenas F, Zevallos S, Kalin RM, Ortlieb L, Fontugne M (2005) Stable isotopes and sclerochronology of the *bivalve Mesodesma donacium*: potential application to Peruvian paleoceanographic reconstructions. Palaeogeogr Palaeoclimatol Palaeoecol 228(1–2):4–25

Carré M, Bentaleb I, Bruguier O, Ordinola E, Barrett NT, Fontugne M (2006) Calcification rate influence on trace element concentrations in aragonitic bivalve shells: evidences and mechanisms. Acta Geochim Cosmochim 70(19):4906–4920

Carstensen D, Arntz W, Schiel S, Held C, Laudien J (2006) Ecology, aquaculture-demands and genetic variability of Chilean and Peruvian surf clams (*Donax marincovichi* and *Mesodesma donacium*). CENSOR Midterm Symposium, Concepción, Bellavista Club, Chile, 4–8 Sept 2006

Carter JG, Campbell DC, With the contribution of C.R. Altaba et al (2011) A synoptical classification of the Bivalvia (Mollusca). Paleontol Contrib 4:1–47

Carvalho YBM, Romano LA, Poersch LHDS (2015) Effect of low salinity on the yellow clam *Mesodesma mactroides*. Braz J Biol 75(1):8–12

Cassista MC, Hart MW (2007) Spatial and temporal genetic homogeneity in the Arctic surfclam (*Mactromeris polynyma*). Mar Biol 152:569–579

Castellanos ZJAd (1948) Estudio Anatómico sobre *Mesodesma mactroides* Desh (Almeja Amarilla). DAGI Publicaciones Técnicas 5(1):1–49

Castellanos ZJAd (1970) Catálogo de los moluscos marinos bonaerenses. Anales de la Comision de Investigaciones Cientificas de la provincia de Buenos Aires 8:9–365

Cauquoin M (1967) Campagne de la Calypso au large des cotes Atlantiques de l'Amérique du Sud (1961–1962) Premiere Partie. 23. Mollusques Lamellibranches: Mactridae. Résultats Scientifiques des Campagnes de la Calypso 8:223–226

Cauquoin M (1969) Mollusques recoltes par M. Hoffstett sur les cotes de l'Equateur et aux Iles Galapagos. Mactridae. Bulletin du Muesum National d'Histoire Naturelle 40:1019–1030

Cauquoin M (1969b) Sur une Nouvelle espece de Mactridae du Bresil. Bulletin du Muesum National d'Histoire Naturelle 40:1175–1178

Cerrato RM, Keith DL (1992) Age structure, growth, and morphometric variations in the Atlantic surf clam, *Spisula solidissima*, from estuarine and inshore waters. Mar Biol 114(4):581–593

Cessine RS, Strand IE (1978) The surf clam as a nonrenewable resource. University of Maryland, Agricultural Experimental Station, Maryland, Contributions 5417 [Miscellaneous Publications 925], pp 1–25

Chamberlin JL (1952) Trans-Panamic distributions of the Mactridae. Annual Report for American Malacological Union, pp 10–12

Chamberlin JL, Stearns F (1963) A geographic study of the clam, *Spisula polynyma* (Stimpson), Serial atlas of the marine environment, vol 3. American Geographical Society, New York, pp 1–12

Chanley PE (1965) Larval development of the brackish water mactrid clam, *Rangia cuneata*. Chesap Sci 6(4):209–213

Chavan A (1969) Superfamily Lucinacea. In L. R. Cox et al., eds., Part N [Bivalvia], Mollusca 6, vols. 1 and 2: xxxvii + 952 p. In R. C. Moore, ed., Treatise on invertebrate paleontology. Geological Society of America and University of Kansas Press, Lawrence, Kansas. pp N491–N517

Chemnitz JHM (1782) Neues systematisches Conchylien-Cabinet/Geordnet und beschrieben von Friedrich Heinrich Wilhelm Martini und unter dessen Aufsicht nach der Natur gezeichnet und mit lebendigen Farben erleuchtet, vol 6. Bey Gabriel Nikolaus Raspe, Nuremberg

Chemnitz JHM (1788) Neues systematisches Conchylien-Cabinet / Geordnet und beschrieben von Friedrich Heinrich Wilhelm Martini und unter dessen Aufsicht nach der Natur gezeichnet und mit lebendigen Farben erleuchtet, vol 10. Bey Gabriel Nikolaus Raspe, Nuremberg

Chenu JC (1859/1862). Manuel de Conchyliologie et de Paléontologie Conchyliologique. Masson, Paris (vol 1, pp 1–508 (1859); vol 2, pp 1–327 (1862))

Children JG (1822) [Anonymous] Art. VI. – Lamarck's genera of shells translated from French by J. G. Children, F. R. S. &c. &c. &c. with plates from original drawings, by Miss Anna Children. Quarterly J of Sci, Lit and Arts 14(27):64–86

Cintra-Buenrostro, C.E., K.W. Flessa and D.L. Dettman. 2012. Restoration flows for the Colorado River estuary, México: estimates from oxygen isotopes in the bivalve mollusk *Mulinia coloradoensis* (Mactridae: Bivalvia). Wetlands Ecol Manag 20(4): 313-327

Clark BL (1922) A new family and new genus from the Tertiary of the Pacific coast. Bull Dept Geol Sci Univ Calif Publ 14(4):115–122

Clark P (1973) The growth rate of the juvenile gaper clam, *Tresus nuttallii*, of Elkhorn Slough, California [abs]. Echo 5:24

Clark P, Nybakken JW, Laurent L (1975) Aspects of the life history of *Tresus nuttallii* in Elkhorn Slough. Calif Fish Game 61(4):215–227

Cledón M, Nuñez JD (2010) Siphon nipping facilitates lethal predation in the clam *Mesodesma mactroides* (Reeve, 1854) (Mollusca: Bivalva). Mar Biol 157(4):737–745

Climo FM (1972) The reinstatement of *Mactra murchisoni* Deshayes, 1854 (Mollusca: Bivalvia: Mactridae). Records Dominion Mus 6(5):99–105

Closs D, Forti Esteves IR (1971) Quaternary mollusks from the Santa do Plamar County. Iheringia 4:19–58

Clotteau G, Dubé F (1993) Optimization of fertilization parameters for rearing surf clams (Spisula solidissima). Aquaculture 114(3–4):339–353

Coan EV, Kabat AR (2017) The malacological contributions of Rudolph Amandus Philippi (1808–1904). Malacologia 60:31–323

Coan EV, Petit RE (2011) The publications and malacological taxa of William Wood (1774–1857). Malacologia 54:1–76

Coan EV, Valentich Scott P (1997) Checklist of the marine bivalves of the northeastern Pacific Ocean. Museum of Natural History, Santa Barbara

Coan EV, Valentich-Scott P (2007) Bivalvia. In: Carlton JT (ed) The Light and Smith manual: intertidal invertebrates from central California to Oregon, 4th edn, completely revised and expanded edn. University of California Press, California, pp 807––859

Coan EV, Valentich-Scott P (2012) Bivalve seashells of tropical West America. Marine Bivalve Mollusks from Baja California to Perú. Santa Barbara Museum of Natural History monographs, Number 6, Studies in biodiversity, Number 4, 2 vols, Santa Barbara Museum of Natural History, Santa Barbara, CA

Coan EV, Valentich-Scott P, Bernard FR (2000) Bivalve seashells of western North America: marine bivalve mollusks from Arctic Alaska to Baja California. Santa Barbara Museum of

Natural History monographs, Number 2; Studies in biodiversity, Number 2. Santa Barbara Museum of Natural History, Santa Barbara, CA

Coch NK (1971) Geology of the Newport News South and Bowers Hill quadrangles, Virginia, vol 28. Virginia Division of Mineral Resources Report of Investigations, pp 1–26

Coelho PA, Torres SN, Ramírez WE, Gutiérrez PA, Toro CA, Soto JG, Sbarbara DG, Pezoa JE (2016) A machine vision system for automatic detection of parasites *Edotea magellanica* in shell-off cooked clam *Mulinia edulis*. J Food Eng 181:84–91

Conrad TA (1831) Description of fifteen new species of recent and three of fossil shells, chiefly from the coast of the United States. J Acad Natl Sci Phila 6:256–268

Conrad TA (1832) American Marine Conchology. T. A. Conrad, Philadelphia. 72 pp (published in parts: 1: 1–12, pls. 1, 2, April 1831; 2: 13–28, pls. 3–5, September, 1831; 3: 29–40, pls. 6–8, May, 1832; 4: 41-?, pls. 9-?, 1832; 5: ?-71, pls. ?-17, 1833)

Conrad TA (1833) On some new fossil and recent shells of the United States. Am J Sci Arts 23:339–346

Conrad TA (1837) Description of the marine shells from upper California, collected by Thomas Nuttall. J Acad Natl Sci Phila 7:227–268

Conrad TA (1838) Fossils of the Medial Tertiary of the United States: illustrated by figures drawn from nature. J. Dobson, Philadelphia. Part no. 1, pp i–xvi, 1–32, pls 1–17

Conrad TA (1840) On the Silurian system, with a table of the strata and characteristic fossils. Am J Sci Arts 38:86–93

Conrad TA (1845) Fossils of the Miocene formation of the United States: illustrated by figures drawn from nature. J. Dobson, Philadelphia. Part no. 3, pp 57–80, pls 30–32, 34–44

Conrad TA (1853) Notes on shells, with descriptions of new species. Proc Acad Natl Sci Phila 6(7):199–200

Conrad TA (1857) Report on the paleontology of the survey. In: Thomas Antisell, geological report [Parke's surveys in California and Arizona]. United States Pacific Railroad Exploring expedition Reports 7(2):189–196

Conrad TA (1860) Descriptions of new species of cretaceous and Eocene fossils of Mississippi and Alabama. J Acad Nat Sci Phil 2(4):275–298

Conrad TA (1862) Catalogue of the Miocene shells of the Atlantic slope Proc Acad Nat Sci of Phil 14: 559–586

Conrad TA (1867) Description of new West coast shells. Am J Conchol 3(2):192–193

Conrad TA (1868) Catalogue of the family Mactridae. Am J Conchol 3(3). appendix 30–47

Conroy AM, Smith PJ, Michael KP, Stotter DR (1993) Identification and recruitment patterns of juvenile surf clams, *Mactra discors* and *M. murchisoni* from central New Zealand. N Z J Mar Freshw Res 27(3):279–285

Cooke AH (1895) Molluscs, Cambridge natural history, vol 3. MacMillan, London

Cooke CW (1936) Geology of the coastal plain of South Carolina. U S Geol Surv Bull 837:1–196

Cooley NR (1978) An inventory of the estuarine fauna in the vicinity of Pensacola, Florida, Florida Marine Research Publications, vol 31. Marine Research Laboratory, Florida, pp 1–119

Cooper RB (1981) Salinity tolerance of *Rangia cuneata* (Pelecypoda: Mactridae) in relation to its estuarine environment: a review. Walkerana 1:19–31

Coscaron S (1959) La almeja amarilla (*Mesodesma* (*T.*) *mactroides* Deshayes) de la costa de la provincia de Buenos Aires. Agro Publicacion Tecnica 1(3):1–66

Cox LR (1969) General features of Bivalvia. In: Moore RC (ed) Treatise on invertebrate paleontology. Kansas Geological Society of America and University of Kansas, Lawrence, pp 2–129

Craeymeersch JA, Faasse MA, Gheerardyn H, Troost K, Nijland R, Engelberts A, Perdon KJ, Van den Ende D, Van Zwol J (2019) First records of the dwarf surf clam Mulinia lateralis (Say, 1822) in Europe. Mar Biodiver Rec 12(5):1–11

Cremonte F (2001) *Bartolius pierrei* ng, n. sp. (Digenea: Gymnophallidae) from the Península Valdés, Argentina. Syst Parasitol 49(2):139–147

Cremonte F, Figueras A (2004) Parasites as possible cause of mass mortalities of the presently critically endangered clam *Mesodesma mactroides* on the Southwest Atlantic coast. Bull Eur Assoc Fish Pathol 24(3):166–171

Cremonte F, Ituarte C (2003) Pathologies elicited by the gymnophallid metacercariae of *Bartolius pierrei* in the clam *Darina solenoides*. J Mar Biol Assoc U K 83(2):311–318

Cripe GM (2006) Contaminated sediment testing with the bivalve Mulinia lateralis: culture refinement for organism availability. Environ Toxicol Chem 25(5):1332–1336

Cruz M (1983) Bivalvos del golfo de Guayaquil. Acta Oceanográfica del Pacífico 2(2):735–821

Cuvier GLCFD (1798) Tableau Élémentaire de l'Histoire Naturelle des Animaux. Baudouin, Paris

D'Amico VL, Hernández MA, Bala LO (2004) Selección de presas en relación con las estrategias de forrajeo de aves migratorias en Península Valdés, Argentina. Ornitol Neotropical 15:357–364

d'Orbigny AD (1834–1847) Mollusques. In: Bertrand CP (ed) Voyage dans l'Amerique Meridionale (Le Bresil, La Republique Orientale de L'Uruguay, La Republique Argentine, La Patagonie, La Republique du Chili, La Republique de Bolivia, La Republique du Perou), execute pendant les annees 1826, 1827, 1828, 1829, 1830, 1831, 1832 et 1833. Chez Ve. Levrault, Paris

Dadon JR (1998) Estado del Recurso Almeja Amarilla (*Mesodesma mactroides*) en el Municipio de La Costa durante el período diciembre 1997-abril 1998. Informe Técnico Convenio. Facultad de Ciencias Exactas y Naturales, Universidad de Buenos Aires – Municipio de La Costa, Buenos Aires

Dadon JR (1999a) Evaluación del Recurso Almeja Amarilla (*Mesodesma mactroides*) en el Partido de La Costa durante el año 1998. Informe Técnico Convenio. Facultad de Ciencias Exactas y Naturales, Universidad de Buenos Aires – Municipio de La Costa, Buenos Aires

Dadon JR (1999b) Variaciones Anuales de la Pesquería de Almeja Amarilla (*Mesodesma mactroides*) en el Noreste de la Provincia de Buenos Aires (Argentina). Resultados del XIV Simposio Científico-Tecnológico. Comisión Técnica Mixta del Frente Marítimo, Montevideo, Uruguay, pp 59–60

Dadon JR, Castaños C (1995) Evaluación de efectivos de la almeja *Mesodesma mactroides* en el sector norte de la costa bonaerense, Octubre 1994–Marzo 1995 argentina. Resultados del VI Congreso Latinoamericano de Ciencias del Mar, Mar del Plata, Argentina

Dadon JR, Chiappini GMT (2000) La almeja amarilla (*Mesodesma mactroides*) en el Partido de La Costa durante el año 1999. Informe Técnico Convenio. Facultad de Ciencias Exactas y Naturales, UBA, Universidad de Buenos Aires, Municipio de La Costa, Buenos Aires

Dadon JR, Castaños C, García PE, Chiappini GMT, Cruses JM (1997) Aplicación de un Modelo Ambiental al estudio de *Mesodesma mactroides* y *Donax hanleyanus* en la Costa Septentrional de Buenos Aires, Argentina. Resultados expandidos, VII Congreso Latinoamericano de Ciencias del Mar, San Pablo, Brasil, vol 1, pp 228–229

Dadon JR, Castaños C, Pérez García RE, Chiappini GMT, Cruses JM (2001) Efectos a largo plazo de las pesquerías de almeja amarillas (*Mesodesma mactroides*) y berberecho (*Donax hanleyanus*), y de la urbanización sobre las comunidades intermareales. In: Krisler Alveal V, Tarsicio AJ (eds) Sustentabilidad de la biodiversidad, un problema actual. Bases científico-técnicas, teorizaciones y proyecciones. Universidad de Concepción, Chile

Daley GM (2002) Creating a paleoecological framework for evolutionary and paleoecological studies: an example from the Fort Thompson Formation (Pleistocene) of Florida. PALAIOS 17(5):419–434

Dall WH (1889) A preliminary catalogue of the shell-bearing marine mollusks and brachiopods of the southeastern coast of the United States, with illustrations of many of the species. Bull U S Natl Mus 37:1–121

Dall WH (1891) On some marine mollusks from the southern coast of Brazil. Nautilus 5:42–45

Dall WH (1894a) On the species of *Mactra* from California. Nautilus 7(12):136–138

Dall WH (1894b) On some species of *Mulinia* from the Pacific coast. Nautilus 8(1):5–6

Dall WH (1894c) Synopsis of the Mactridae of North America. Nautilus 8(3):25–28

Dall WH (1894d) Synopsis of the Mactridae of Northwest America, south to Panama. Nautilus 8(4):39–43

Dall WH (1894e) Monographs of the genus *Gnathodon* Gray (*Rangia*, Desmoulins). Proc U S Natl Mus 17:89–107

Dall WH (1895a) A new classification of the Pelecypoda. Transactions of the Wagner Free Institute of Science of Philadelphia, vol 3, Tertiary Mollusks of Florida. Contributions to the Tertiary fauna of Florida, with especial reference to the Miocene silex-beds of Tampa and the Pliocene beds of the Caloosahatchie River. Part III. pp 481–483, 484–570

Dall WH (1895b) Synopsis of a review of the genera of recent and Terciary Mactridae and Mesodesmatidae. Proc Malacol Soc Lond 1:203–213

Dall WH (1897) List of species collected at Bahia, Brazil, By Dr. H. Von Ihering. Nautilus 10:121–132

Dall WH (1898a) Contributions to the Tertiary fauna of Florida with especial reference to the Miocene Silex beds of Tampa and the Pliocene beds of the Caloosahatchie River, including in many cases a complete revision of the generic groups treated and of their American Tertiary Species. Trans Wagner Free Inst Sci Phila 3:571 947

Dall WH (1898b) Note on the anatomy of *Resania*, Gray, and *Zenatia*, Gray. Proc Malacol Soc Lond 3(2):85–86

Dall WH (1901) Result of the Branner–Agassiz expedition to Brazil. Proc Wash Acad Sci 3:139–147

Dall WH (1902) Illustrations and descriptions of new, unfigured or imperfectly known shells, chiefly American, in the U. S. National Museum. Proc U S Nat Mus 24(1264):499–566

Dall WH (1909) Report on a collection of shells from Peru, with a summary of the littoral marine Mollusca of the Peruvian Zoological province. Proc U S Natl Mus 37:147–294

Dall WH (1915) Notes on American species of *Mactrella*. Nautilus 9:61–63

Dall WH (1916a) Checklist of the Recent Bivalve Mollusks (Pelecypoda) of the northwest coast of America from the Polar Sea to San Diego, California. South-west Museum, Los Angeles

Dall WH (1916b) Diagnosis of new species of marine bivalve mollusks from northwest coast of America in the collection of the United States National Museum. Proc U S Natl Mus 52(2183):393–417

Dall WH (1921a) Summary of the marine shellbearing mollusks of the northwest coast of America, from San Diego, California, to the Polar Sea, mostly contained in the collection of the U. S. National Museum, with illustrations of hitherto unfigured species. Bull U S Natl Mus 112:iii + 217

Dall WH (1921b) New shells from the Pliocene or early Pleistocene of San Quentin Bay, Lower California. West Am Scient 19(3):21–23

Dall WH (1925) Illustrations of unfigured types of shells in the collection of the United States National Museum. Proc U S Natl Mus 66:1–41

Darragh TA, Kendrick GW (1971) *Zenatiopsis ultima* sp. nov., terminal species of the *Zenatiopsis* lineage (Bivalvia: Mactridae). Proc R Soc Victoria 84(1):87–92

Dautzenberg P (1921) Contribution a la fauna malacologique du Cameroun. Extrait de la Revue Zoologique Africaine 9:88–191

Davies DJ, Powell EN, Stanton RJ (1989) Taphonomic signature as a function of environmental process: shells and shell beds in a hurricane-influenced inlet on the Texas coast. Palaeogeogr Palaeoclimatol Palaeoecol 72:317–356

Davis JD (1963) A study of the Arctic Wedge Clams, *Mesodesma deauratum* (Turton) and Mesodesma arctatum (Conrad) of the Northwestern Atlantic. PhD Thesis, University of New Hampshire

Davis JD (1964) Lectotype designation for *Mesodesma arctatum*. Nautilus 78(1):3–6

Davis JD (1965) *Mesodesma deauratum*: synonymy, holotype and type locality. Nautilus 78(3):96–100

Davis C, Shumway SE (1996) Larval and juvenile growth of Stimpson's surf clam, a new candidate species of aquaculture development. J Shellfish Res 15(2):479–480

De Castro GA, Dos Santos EF (1989) Levantamento preliminar de moluscos em praias arenosas e areno-lodosas de Piúma, estado do Espírito Santo, Brasil. Memóroas do instituto Oswaldo Cruz 84(4):101–104

De Joannis L (1834) Mésodesme. *Mesodesma*. Deshayes Magasin de Zoologie 5 Classe V [1 pp] pl 54

De Kay JE (1843) Zoology of New York, or the New York Fauna; comprising detailed descriptions of all the animals hitherto observed within the state of New York; with brief notices of

those occasionally found near its borders: and accompanied by appropriate illustrations. Part 5. Mollusca. Carroll and Cook, Printers to the Assembly, Albany, NY
De Moreno JE, Moreno VJ, Brenner RR (1976a) Lipid metabolism of the yellow clam, *Mesodesma mactroides*: I. Composition of the lipids. Lipids 11(4):334
De Moreno JE, Moreno VJ, Brenner RR (1976b) Lipid metabolism of the yellow clam, *Mesodesma mactroides*: 2-polyunsaturated fatty acid metabolism. Lipids 11(7):561
Dean HK (1980) Microstructure of the crystalline style of *Spisula solidissima* (Bivalvia, Mactridae). Nautilus 94:54–58
Defeo O (1985) Aspectos biocenologicos y dinámica de la población de almeja amarilla, *Mesodesma mactroides* (Deshayes, 1854) en la zona de la barra de Chuy, Dpto de Rocha, Uruguay. Contribuciones del Departamento de Oceanografía de la Facultad de Humanidades y Ciencias 2(3):50–74
Defeo O (1989) Development and management of artisanal fishery for yellow clam *Mesodesma mactroides* in Uruguay. Aust Fish 7(3):21–25
Defeo O, Scarabino V (1990) Ecological significance of a possible deposit-feeding strategy in *Mesodesma mactroides* (Deshayes, 1854) (Mollusca: Pelecypoda). Atlantica 12(1):55–65
Defeo O, Layerle C, Masello A (1986) Spatial and temporal structure of the yellow clam *Mesodesma mactroides* (Deshayes 1854) in Uruguay. Medio Ambiente 8:48–57
Defeo O, Ortiz E, Castilla JC (1992) Growth, mortality and recruitment of the yellow clam *Mesodesma mactroides* on Uruguayan beaches. Mar Biol 114:429–437
Degraer S, Meire P, Vincx M (2007) Spacial distribution, population dynamics and productivity of *Spisula subtruncata*: implications for *Spisula* fisheries in seaduck wintering areas. Mar Biol 152(4):863–875
Deshayes GP (1832, 1830a, 1830b, 1832a, 1832b). In Bruguière JG, Deshayes GP, Hwass CH 1789–1832 (eds) Encyclopédie Méthodique. Histoire Naturelle des Vers. Tome 2, Première partie [author Deshayes, publisher Agasse, Paris], pp vii + 1–256, Paris [February 1, 1830a]. Tome 2, Deuxième partie [author Deshayes, publisher Agasse, Paris], pp 1–144 [February 1, 1830b], pp 145–594, including one unnumbered table [September 29, 1832b]. Tome 3 [author Deshayes, publisher Agasse, Paris], pp 595–1152 [September 29, 1832c]
Deshayes GP (1853) [J. Gray, ed.; authorship attributed by gray to Deshayes: "Catalogum elaboravit dominus Deshayes ..."] Catalogus Concharum Bivalvium quae in Museo Britannico Asservantur/Catalogue of the Conchifera or Bivalve shells in the collection of the British Museum. Part 1, Veneridae, Cyprinidae and Glauconomidae. London
Deshayes GP (1854) Description of fourteen new species of Mactra, in the collection of Mr. Cuming. Proc Zool Soc London 21:14–17
Deshayes GP (1855a) Descriptions of new shells from the collection of Hugh Cuming, Esq. Proc Zool Soc London 22:62–72
Deshayes GP (1855b) Descriptions of new shells from the collection of Hugh Cuming, Esq. Proc Zool Soc London 23:317–371
Deshayes GP, Milne-Edwards H (1835) Histoire naturelle des animaux sans vertébres, présentant les charactéres généraux par J.P.B.A. de Lamarck, Histoire des Mollusques, vol 6. J. B. Bailliére, Paris
Des Moulins C (1832) Description d'un nouveau genre de coquille bivalve, fluviatile, de l'Amérique Septentrionale. Soc Linn Bordeaux, Actes 5(26):1–13
DesVoigne DM, Mix MC, Pauley GB (1970) A papillomalike growth on the siphon of the horse clam, *Tresus nuttalli*. J Invertebr Pathol 15(2):262–267
Díaz P, Campos B (2014) Ontogenia de la concha larval y postlarval de cuatro especies de bivalvos de la costa del Pacífico sureste. Rev Biol Mar Oceanogr 49(2):175–191
Díaz O, Encina F, Recabarren E, Del Valle C, Pastene R, Montes S, Figueroa A (2008) Estudio de la concentración de arsénico, mercurio, plomo y fenantreno en la macha (*Mesodesma donacium*): implicancias alimentarias y toxicológicas. Rev Chilena Nutr 35(1):53–60
Diaz Merlano JM, Puyana Hegedus, M (1994) Moluscos del Caribe Colombiano, un Catálogo Ilustrado. Colciencias y Fundación Natura Colombia, INVEMAR, Santefe de Bogotá, Colombia

Dietl GP, Smith JA (2017) Live-dead analysis reveals long-term response of the estuarine bivalve community to water diversions along the Colorado River. Ecol Eng 106(Part B):749–756

Dillwyn LW (1817) A descriptive catalogue of recent shells, arranged according to the Linnean method; with particular attention to the synonymy. Two vols.: vol. 1, i-xii + p. 1–580; vol. 2, p. 581–1092 + 29 p. John and Arthur Arch, London

Dinnel PA (1971) Recruitment, distribution, mortality, and growth of the 1970 and 1971 year classes of the gaper clam, *Tresus Capax* (Gould, 1850) (Bivalvia: Mactridae), in Humboldt Bay, California. Doctoral dissertation, Humboldt State University

Dinnel PA, DeMartini JD (1974) A supposedly extinct bivalve species found living off California. Veliger 17(1):44–46

Doello-Jurado M in Carcelles A (1949) Dos nuevas especies de bivalvos marinos. Comunicaciones Zoologicas del Museo de Historia Natural de Montevideo 3:1–8

DuBar JR (1962) Neogene biostratigraphy of the Charlotte Harbor Area in Southwestern Florida. Florida Geol Surv Geol Bull 43:1–83

Dunker W (1850) Diagnoses molluscorum quorundam novorum. Zeit für Malak 7(2):30–32

Dunker WBRH (1853) Diagnoses molluscorum novorum [part]. Z Malakozool 10(7):110–112

Durham JW (1950) 1940 E. W. Scripps Cruise to the Gulf of California: Part II, Megascopic paleontology and marine stratigraphy. In: Anderson CA, Durham JW, Shepard FP, Natland ML, Revelle R (eds) Geological Society of America Memoir, vol 43. Geological Society of America, Boulder, CO, pp 1–194

Eichwald KEIv (1830) Naturhistorische Skizze von Lithauen, Volhynien and Podolien in Geognostisch-mineralogischer, Botanischer und Zoologischer Hinsicht Entworfen. Joseph Zawadski, Wilna

Emerson WK, Jacobson MK (1976) The American Museum of Natural History Guide to Shells. Land, Freshwater and Marine, from Nova Scotia to Florida. A. A. Knopf, Inc., New York

Engle JB, Merrill AS (1967) The surf clam—New Jersey's most valuable seafood resource. N J Nat News 22(f):148–153

Esqueda-González M, Ríos-Jara E, Galván-Villa CM, Rodríguez-Zaragoza FA (2014) Species composition, richness, and distribution of marine bivalve molluscs in Bahía de Mazatlán, México. ZooKeys 399:43–69

Everson G (1991) Mactra (Mactrotoma) nasuta Gould, 1851. In Bahía Concepción. Festivus 23(9):71–72

Faber MJ (1988) Studies on West Indian marine mollusks. 13. The malacological taxa of Gordon W. Nowell-Usticke. De Kreukel 24(4–5):67–102

Fairbanks LD (1963) Biodemographic studies of the clam *Rangia cuneata* Gray. Tulane Stud Zool 10:3–47

Feder HM, Paul AJ, Paul J (1976) Growth and size-weight relationships of the pinkneck clam *Spisula Polynyma*, in Hartney Bay, Prince William Sound, Alaska. Proc Natl Shellfisheries Assoc 66:21–25

Feder HM, Paul AJ, Paul J (1977) Age and growth of the pinkneck clam, *Spisula polynyma*. Alaska, University of Alaska, Institute of Marine Science, Sea Grant Report 77–12, pp 1–5

Feder HM, Paul AJ, Paul J (1979) The Pinkneck clam, *Spisula polynyma*, in the eastern Bering Sea: growth, mortality, recruitment and size at maturity. Proc Am Assoc Adv Sci Alaska Sci Conf 29:717–738

Félix-Pico EF, Holguin-Quiñones OE, Escamilla-Montes R (2011) Macroinvertebrados marinos asociados al manglar. In: Félix-Pico EF, Serviere E, Riosmena R, León JL (eds) Los manglares de la península de Baja California. Centro Interdisciplinario de Ciencias Marinas, Centro de Investigaciones Biológicas del Noroeste, SC Universidad Autónoma de Baja California Sur, La Paz, Baja California Sur, México, pp 203–232

Ferrari S, Lizarralde Z, Albrieu C, Pittaluga S, Gallardo L (2010) Predación de la almeja *Darina solenoides* por el Ostrero Austral *Haematopus leucopodus* en el estuario del Río Gallegos (Patagonia, Argentina). IV Reunión Binacional de Ecología, Buenos Aires, Agosto 2010

Férussac AEJP, d'Audebard, JF, Deshayes, GP (1819–1851) Histoire Naturelle Générale et Particulière des Mollusques Terrestres et Fluviatiles tant des Espèces que l'on Trouve

Aujourd'hui Vivantes, que des Dépouilles Fossiles de celles qui N'existent Plus: Classes d'après les Caractères Essentiels que Presentment ces Animaux et leurs Coquilles. Baillière, Paris. Published in 42 livraisons

Figueiras A (1961) Contribución al conocimiento de la Malacofauna del Uruguay. Comun Soc Malacol Urug 1(1):15–21

Figueiras A (1962) Sobre nuevos hallazgos de moluscos subfósiles de la Transgresión Querandina. Comun Soc Malacol Urug 1(3):53–68

Figueiras A, Broggi A (1972–1973) Estado actual de nuestros conocimientos sobre los moluscos fósiles del Uruguay. Comun Soc Malacol Urug 3(23–24):203–240

Figueiras A, Sicardi OE (1969) Catálogo de los moluscos marinos del Uruguay parte III. Comun Soc Malacol Urug 2(16–17):355–378

Fiori S, Carcedo MC (2015) Influence of grain size on burrowing and alongshore distribution of the yellow clam (*Amarilladesma mactroides*). J Shellfish Res 34(3):785–790

Fiori S, Cazzaniga N (1999) Mass mortality of the yellow clam, *Mesodesma mactroides* (Bivalvia: Mactracea) in Monte Hermoso beach, Argentina. Biol Conserv 89(3):305–309

Fiori S, Defeo O (2006) Biogeographic patterns in life-history traits of the yellow clam, *Mesodesma mactroides*, in sandy beaches of South America. J Coast Res 22(4):872–880

Fiori S, Morsán E (2004) Age and individual growth of *Mesodesma mactroides* (Bivalvia) in the southernmost range of its distribution. J Mar Sci 61:1253–1259

Fiori S, Cazzaniga N, Estebenet AL (2004a) Winter distribution, density and size of *Mesodesma mactroides* (Bivalvia, Mactracea) in Monte Hermoso beach (Argentina). Braz J Oceanogr 52(1):1–9

Fiori S, Vidal-Martinez VM, Sima-Alvarez R, Rodriguez-Canul R, Aguirre-Macedo ML, Defeo O (2004b) Field and laboratory observations of the mass mortality of the yellow clam *Mesodesma mactroides* in South America: the case of Isla del Jabalí, Argentina. J Shellfish Res 23(2):451–456

Fiori S, Frontini R, Bayón C (2019) Morphometric analysis of the yellow clam (*Amarilladesma mactroides*) to interpret fragmentary archaeological specimens. J Island Coastal Archaeol 14:1–10. https://doi.org/10.1080/15564894.2018.1521482

Fischer P (1880–1887) Manuel de conchyliologie et de paleontologie conchyliologique ou historie naturelle des mollusques vivants et fossiles (vol 1, pp 1–112, 1880; vol 2, pp 113–192, 1881; vol 3, pp 193–304, 1881; vol 4, pp 305–416, 1882; vol 5, pp 417–512, 1883; vol 6, pp 513–608, 1883; vol 7, pp 609–688, 1884; vol 8, pp 689–784, 1884; vol 9, pp 785–896, 1885; vol 10, pp 897–1008, 1886; vol 11, pp 1009–1369, 1887). F. Savy, Paris

Fischer R (1915) Über die Anatomie von *Mactra* (*Mulinia*) *coquimbana* Philippi. Jenaische Zeitschrift Für Naturwissenschaft 53[(n.f.)46](4):597–662

Fischer PH (1977) Attitude de locomotion chez *Mactra lurida* Phil. J Conchyliologie 114[(4)67](3–4):129–130

Fleming J (1818) Conchology. In: Supplement to the Fourth, Fifth, and Sixth editions of the Encyclopaedia Britannica. With Preliminary Dissertations on the History of the Sciences. Illustrated by Engravings, vol 3. Archibald Constable and Company, Edinburgh; Hurst, Robinson and Company, London

Flessa KW (2002) Conservation paleobiology. Am Paleontologist 10(1):2–5

Flessa KW, Rodriguez CA (1999) Conservation implications of the population decline of the Colorado Delta bivalve mollusk, *Mulinia coloradoensis*. Report submitted to the Southwest Center for Biological Diversity

Flessa KW, Téllez-Duarte MA (2001) Taxonomic status and distribution of the bivalve mollusks *Mulinia coloradoensis* in the Gulf of California. Report to the Center for Biological Diversity and Defenders of Wildlife

Foltz DW, Sarver SK, Hrincevich AW (1995) Genetic structure of brackish water clams (*Rangia* spp.). Biochem Syst Ecol 23(3):223–233

Frey HW (1971) California's living marine resources and their utilization. California Department Fish and Game, California

Fritz LW, Ragone LM, Lutz RA (1990) Microstructure of the outer shell layer of *Rangia cuneata* (Sowerby, 1831) from the Delaware River: applications in the studies of population dynamics. J Shellfish Res 9(1):205–213

Frizzell DL (1930) *Schizothaerus nuttallii capax* Gld. not *maxima* Midd. Nautilus 44(2):69

Gabb WM (1866) Palaeontology of California. Vol 2. Geological Survey of California, Palaeontology, Cretaceous and Tertiary Fossils

Gabb WM (1869) Descriptions of new species of South American fossils. Am J Conchol 5(1):25–32

Gabb WM (1873) Article IV, on the topography and geology of Santo Domingo. Trans Amer Phil Soc 15(1):49–259

Gallagher JN, Wells HW (1969) Northern range extension and winter mortality of *Rangia cuneata*. Nautilus 83(1):22–25

Gallardo LJ, Lizarralde ZI (2014) Variación temporal de la población de la almeja *Darina solenoides* y del mejillón *Mytilus edulis platensis*, y su relación con la alimentación del ostrero austral. Informes Científicos-Técnicos UNPA 4(2):44–81

Gallardo-Escárate C, Valenzuela-Muñoz V, Núñez-Acuña G, Haye P (2015) SNP discovery and gene annotation in the surf clam *Mesodesma donacium*. Aquac Res 46(5):1175–1187

Ganong WF (1889) On the economic Mollusca of Acadia. Bull Nat Hist Soc New Brunswick 8:1–116

Gardner J, Mansfield WC (1943) Mollusca from the Miocene and lower Pliocene of Virginia and North Carolina, Part 1, Pelecypoda, with a summary of the stratigraphy. Geol Surv Prof Pap 199-A:1–178

Garrido MV, Chaparro OR, Thompson RJ, Garido O, Navarro JM (2012) Particle sorting and formation and elimination of pseudofaeces in the bivalves *Mulinia edulis* (siphonate) and *Mytilus chilensis* (asiphonate). Mar Biol 159(5):987–1000

Gemmell J, Myers BW, Hertz CM (1987) A faunal study of the bivalves of San Felipe and Environs, Gulf of California: from the Gemmell Collection (1965 to 1976). Festivus 18:1–72

Giacomin M, Jorge MB, Bianchini A (2014) Effects of copper exposure on the energy metabolism in juveniles of the marine clam *Mesodesma mactroides*. Aquat Toxicol 152:30–37

Giberto DA, Bremec C, Acha EM, Mianzan H (2004) Large-scale spatial patterns of benthic assemblages in the SW Atlantic: the Rio de la Plata estuary and adjacent shelf waters. Estuarine Coastal Shelf Sci 61:1–13

Gilardoni C, Carballo MC, Cremonte F (2013) The life cycle and geographical distribution of the monorchiid *Proctotrema bartolii* (Digenea) in the clam *Darina solenoides* from the Patagonian coast, Argentina. J Helminthol 87(4):392–399

Gill ED, Darragh TA (1963) Evolution of the Zenatiinae (Mactridae: Lamellibranchiata). Proc R Soc Victoria 77(1):177–190

Glen W (1959) Pliocene and lower Pleistocene of the western part of the San Francisco Peninsula. Calif Univ Publ Geol Sci 36(2):147–198

Gmelin JF (1791) Caroli a Linné, Systema Naturae per Regna Tria Naturae, Secundum Classes, Ordines, Genera, Species, cum characteribus, differentiis, synonymis, locis. (Editio Decima Tertia [13th edn], Aucta, Reformata ed.)

Godwin WF (1968) The distribution and density of the brackish water clam, *Rangia cuneata*, in the Altamaha River, Georgia. Georgia Department of Natural Resources, Contribution Series, Number 5

Goldfuss GA (1820) Handbuch der Zoologie. In G. H. Schubert, Handbuch der Naturgeschichte zum Gebrauch bei Vorlesungen, Theil 3, Abtheilung 1. Schrag, Nürnberg

Goldfuss GA (1841) In Goldfuss GA (1833, 1835, 1836, 1837, 1841). Petrefacta Germaniae tam ea, quae in Museo Universitatis Regiae Borussicae Fridericiae Wilhelmiae Rhenanae Servantur quam alia Quaecunquae in Museis Hoenighusiano Muensteriano Aliisque Extant, Iconibus et Descriptionibus Illustrata. Abbildungen und Beschreibungen der Petrefacten Deutschlands und der Angrenzender Länder unter Mitwirkung des Herrn Grafen Georg zu Münster. Arnz & Co., Düsseldorf. 3 text vols. + atlas. Erster Theil, [vii] + [1] + p. 2–252 ["1826–1833"], with Atlas, pls. 1–71 [1826–1833]. Zweiter Theil [Bivalvia], ["1834–1840", = 1833–1841], part 1, [i] + p. 1–68, with Atlas, pls. 72–96 [1833]; part 2, p. 69–140, with Atlas. pls. 97–121 [1835]; part

3, p. 141–224, with Atlas, pls. 122–146 [1837]; part 4, p. 225–312, with Atlas, pls. 147–165 [1841]
Goldberg R (1989) Biology and culture of the surf clam. In: Manzi JJ, Castagna M (eds) Clam culture in North America. Elsevier, Amsterdam, pp 263–276
Gondim AI, Dias TLP, Campos FF, Alonso C, Christoffersen ML (2011) Benthic macrofauna from Areia Vermelha Marine State Park, Cabedelo, Paraíba, Brazil. Biota Neotropica 11(2):75–86
Gonzalez C, Pía Mena C, Gallardo MH, Clasing E (2002) Variabilidad genética en los miembros del género *Mulinia* (Gray, 1837) (Bivalvia, Mactridae) a lo largo de la costa chilena. Cienc Tecnol Mar 25:147–156
Goodger TE (1972) The reproductive system of the gaper or horse clam, *Tresus Capax* (Pelecypoda: Mactirdae): its anatomical development and histology. Doctoral dissertation, Humboldt State University
Gould AA (1841) Report on the invertebrata of Massachusetts, comprising the Mollusca, Crustacea, Annelida, and Radiata. Folsom et al, Cambridge, MA
Gould AA (1850) Shells collected by the U.S. Exploring Expedition under the command of Charles Wilkes. Proc Boston Soc Nat Hist 3(3):214–218
Gould AA (1851) Descriptions of California shells. Proc Boston Soc Nat Hist 4:87–93
Gould AA (1852–1861) Atlas. Mollusca and shells. In: United States Exploring Expedition. During the years 1838, 1839, 1840, 1841, 1842. Under the command of Charles Wilkes. U. S. N. C. Sherman and Son, Philadelphia
Gould AA (1870) Report on the invertebrata of Massachusetts, published agreeably to an order of the legislature. In: Binney WG (ed) Comprising the Mollusca, 2nd edn. Wright and Porter, Boston
Grant US, Gale HR (1931) Catalogue of the marine Pliocene and Pleistocene Mollusca of California and adjacent regions. Mem San Diego Soc Nat Hist 1:1–1036
Gray JE (1825) A list and description of some species of shells not taken notice of by Lamarck. Ann Philos 9:134–140
Gray JE (1837) A synoptical catalogue of the species of certain tribes of genera of shells contained in the collection of the British Museum. Mag Nat Hist 1:370–376
Gray JE (1840) Shells of molluscous animals. In: Synopsis of the contents of the British Museum, 42nd edn. G. Woodfall and Son, London, pp 86–89, 105–152
Gray JE (1847) A list of the genera of recent mollusca, their synonyma and types. Proc Zool Soc Lond 1847:129–219
Gray JE (1853) A revision of the genera of some of the families of conchifera or bivalve shells. Ann Nat Hist II 11:33–44
Griffith E, Pidgeon E (1833–1834) The mollusca and radiata. In: Griffith E (ed) The animal kingdom arranged in conformity with its organization, by the Baron Cuvier, member of the Institute of France, andc. andc. andc. with supplementary additions to each order, by Edward Griffith, F.L.S., A.S., corresponding member of the Academy of Natural Sciences of Philadelphia, and C. and others, vol 12. Whittaker and Co, London
Güller M, Zelaya D (2015) Bivalves from the Chilean Fjords region: knowns and unknowns. Am Malacol Bull 33:233–245
Gundlach J (1883a) Apuntes para la fauna Puerto-Riqueña. Anales de la Sociedad Española de Historia Natural, Cuarta Parte A – Moluscos Terrestres y Fluviatiles 12:1–58
Gundlach J (1883b) Apuntes para la fauna Puerto-Riqueña. Anales de la Sociedad Española de Historia Natural, Quinta Parte B – Moluscos marinos 12:411–484
Guppy RJL (1875) On new species of bivalve Mollusca found at Cumana, Venezuela. Ann Mag Nat Hist 4(15):49–50
Habe T (1979) *Mactra bonneaui* Bernardi, a synonym of *Mactra veneriformis* Reeve. Venus 38(1):66–67
Haderlie EC, Abbott DP (1980) Bivalvia: the clams and allies. In: Morris RH, Abbott DP, Haderlie EC (eds) Intertidal invertebrates of California. Stanford University Press, California, pp 355–411

Hanley SCT (1842–1856) An illustratted and descriptive catalogue of recent Bivalve shells. [With 960 figures by Wood and Sowerby, forming an appendix to the Index Testaceologicus]. Williams and Norgate, London

Harbo RM, McDaniel N, Swanston D, LaFollette P (2012) An exciting new discovery: the lightly-sculptured odostome snail, *Evalea tenuisculpta* (Carpenter, 1864) feeding on the siphon tips of the fat gaper, *Tresus capax* (Gould, 1850) in Vancouver Harbor, British Columbia. Dredgings 52(2):3–4

Hare MP, Weinberg JR (2005) Phylogeography of surfclams, *Spisula solidissima*, in the western North Atlantic based on mitochondrial and nuclear DNA sequences. Mar Biol 146(4):707–716

Hare MP, Weinberg JR, Peterfalvy O, Davidson M (2010) The southern surfclam (*Spisula Solidissima similis*) found north of its reported range: a commercially harvested population in Long Island Sound, New York. J Shellfish Res 29(4):799–807

Harry HW (1969) Anatomical notes on the Mactrid bivalve, *Raeta plicatella* Lamarck, 1818, with a review of the genus *Raeta* and related genera. Veliger 12:1–23

Heiman EL, Mienis HK (2010) Shells of east Sinai, an illustrated list. Family Mactridae. Triton 22:2–3

Henderson J (1931) *Schizothaerus nuttallii* and varieties on Puget Sound. Nautilus 45(1):32–33

Hendy AJ, Jones DS, Moreno F, Zapata V, Jaramillo C (2015) Neogene molluscs, shallow marine paleoenvironments, and chronostratigraphy of the Guajira Peninsula, Colombia. Swiss J Palaeontol 134(1):45–75

Hernández-Alvarez A, Carmona R, Arce N (2013) Feeding ecology of Red Knots *Calidris canutus roselaari* at Golfo de Santa Clara, Sonora, Mexico. Wader Study Group Bull 120(3):194–201

Herrmann M, Alfaya JE, Lepore M, Penchaszadeh PE, Laudien J (2009) Reproductive cycle and gonad development of the Northern Argentinean *Mesodesma mactroides* (Bivalvia: Mesodesmatidae). Helgol Mar Res 63(3):207–218

Herrmann M, Alfaya JE, Lepore M, Penchaszadeh PE, Arntz WE (2011) Population structure, growth and production of the yellow clam *Mesodesma mactroides* (Bivalvia: Mesodesmatidae) from a high-energy, temperate beach in northern Argentina. Helgol Mar Res 65(3):285–297

Hershler R, Jayko A (2009) A mactrid bivalve from Pleistocene deposits of lake Russell, Mono basin, California. J Paleontol 83:496–499

Hertlein LG, Grant US IV (1972) The geology and paleontology of the marine Pliocene of San Diego, California (Paleontology: Pelecypoda). San Diego Soc Nat Hist Mem 2:135–409

Hertlein LG, Strong AM (1950) Eastern Pacific expeditions of the New York Zoological Society. XLII. Mollusks from the west coast of Mexico and Central America. Part IX. Zoologica 35(4):217–252

Hertlein LG, Strong AM (1955) Marine mollusks collected during the Askoy expeditions to Panama, Columbia, and Ecuador in 1941. Bull Am Mus Nat Hist 107:159–318

Hetherington R, Barrie JV, Reid RG, MacLeod R, Smith DJ, James TS, Kung R (2003) Late Pleistocene coastal paleogeography of the Queen Charlotte Islands, British Columbia, Canada, and its implications for terrestrial biogeography and early postglacial human occupation. Can J Earth Sci 40(12):1755–1766

Hickman CS (2015) Paleogene marine bivalves of the deep-water Keasey Formation in Oregon, Part III: The heteroconchs. PaleoBios 32(1):1–44

Hill FC (1983) Unexplained occurrence of the mactrid bivalve *Rangia cuneata*, from the Arrowhead Farms Indian Site near Louisville, Kentucky. Nautilus 97(2):79–81

Hodge JT (1841) ART. VIII. Observations on the secondary and tertiary formation. Am J Sci Arts 41(2):332–348

Hodson F, Hodson HK (1931) Venezuelan mollusks. Bull Am Paleontol 16(59):1–132

Holm GP (2008) A third *Tresus* found in Pacific Northwest. Dredgings 48(2):8

Holme NA (1959) The British species of *Lutraria* (Lamellibranchia), with a description of *L. angustior* Philippi. J Mar Biol Assoc U K 58(3):557–568

Holmes FS (1860) Post-Pleiocene fossils of South-Carolina. Russell & Jones, Charleston, SC

Hopkins SH, Andrews JD (1970) *Rangia cuneata* on the East Coast: thousand-mile range extension, or resurgence? Science 167(3919):868–869

Hopkins SH, Anderson JW, Horvath K (1974) Biology of the clam *Rangia cuneata*: what we know and what it means. Proc Natl Shellfisheries Assoc 64:1–4

Huang YM, Wang FP (1995) Studies on the morphology of *Mactra antiquata* Spengler. Trans Chin Soc Malacol 5–6:50–60

Huber M (2010) Compendium of bivalves. A full-color guide to 3,300 of the World's Marine Bivalves. A status on Bivalvia after 250 years of research. ConchBooks, Hackenheim, Germany

Huber M (2015) Compendium of bivalves 2. A full-color guide to the remaining seven families. A systematic listing of 8500 Bivalve species and 10500 synonyms. ConchBooks, Harxheim, Germany

Hupé LH (1854) Mollusca. In: Gay C (ed) Historia Fisica y Politica de Chile. Paris, Maulde y Renou

Hutcheson MS, Stewart PL (1994) A possible relict population of *Mesodesma deauratum* (Turton): Bivalvia (Mesodesmatidae) from the Southeast Shoal, Grand Banks of Newfoundland. Can J Fish Aquat Sci 51(5):1162–1168

Hutton FW (1873) Catalogue of the marine mollusca of New Zealand, with diagnoses of new species. Colonial Museum and Geological Survey, Wellington, New Zealand

Hylleberg KJ, Knudsen J, Vongpanich V (1999) The species described by Lorentz Spengler in his "Beskrivelse over det toskallede Conchylie-Slaegt *Mactra*". A translation into English. Phuket Mar Biol Center Spec Publ 21(2):499–518

ICZN (1944) On the status of names first published in volumes 1 to 11 of Martini (F. H. W.) and Chemnitz (J. H.), Neues Systematisches Conchylien–Cabinet. Nürenberg, 1769–1795. Opin Declar Rend Int Comm Zool Nomen 3:25–36

Imai T, Hatanaka M, Sato R, Sakai S (1953) Tank breeding of the Japanese surf clam, *Mactra sachaliensis* Schrenck. Sci Rep Res Inst Tohoku Univ 4:121–131

Iredale T (1930) More notes on the marine Mollusca of New South Wales. Rec Aust Mus 17:384–407

Iredale T (1931) Australian molluscan notes. Rec Aust Mus 18:201–235

Iredale T, McMichael DF (1962) A reference list of the marine Mollusca of New South Wales. Aust Mus Mem 11:1–109

Ivanov SV (1980) Stroenie rakovin Molliuskov semeistva Mactridae. [Shell structure of mollusks of the family Mactridae]. Biologiia Moria 6(3):62–66. [translation: Sov J Mar Biol 6(3):159–164]

Ivany LC, Wilkinson BH, Jones DS (2003) Using stable isotopic data to resolve rate and duration of growth throughout ontogeny: an example from the surf clam, Spisula solidissima. PALAIOS 18(2):126–137

Jacobson MK (1972a) Observation on the siphonal behavior of young surf clams, *Spisula solidissima*. Nautilus 86(1):25

Jacobson MK (1972b) On deformations of the surf clam. N Y Shell Club Notes 179:5

Jacobson MK, Old WE Jr (1966) On the identity of *Spisula similis* (Say). Rep Am Malacol Union 1966:30–31

Jacot AP (1920) Some marine molluscan shells of Beaufort and Vicinity. J Elisha Mitchell Sci Soc 36(3/4):129–145

Jaramillo E, Pino M, Filun L, González M (1994) Longshore distribution of *Mesodesma donacium* (Bivalvia: Mesodesmatidae) on a Sandy beach of the south of Chile. Veliger 37(2):192–200

Jeffreys JG (1864) British conchology, or an account of the Mollusca which now inhabit the British Isles and the surrounding seas. Vol. II. Marine shells, comprising the Brachiopoda, and Conchifera from the family Anomiidae to that of the Mactridae. van Voorst, London

Johnson R (1964) The recent mollusca of August Addison Gould. Bull U S Natl Mus 239:1–172

Jonas, JH (1844) Vorläufige Diagnosen neuer Conchylien, welche ausführlicher beschrieben und abgebildet nächstens erscheinen werden. Zeitschrift für Malakozoologie, Jahrgang 1844, March:33–37

Jones DS (1980) Annual cycle of shell growth increment formation in two continental shelf bivalves and its paleoecologic significance. Paleobiology 6(3):331–340

Jones DS (1981) Annual growth increments in shells of *Spisula solidissima* record marine temperature variability. Science 211(4478):165–167

Jones DS, Thompson I, Ambrose W (1978) Age and growth rate determinations for the Atlantic surf clam *Spisula solidissima* (Bivalvia: Mactracea), based on internal growth lines in shell cross-sections. Mar Biol 47(1):63–70

Jones DS, Williams DF, Arthur MA (1983) Growth history and ecology of the Atlantic surf clam, *Spisula solidissima* (Dillwyn), as revealed by stable isotopes and annual shell increments. J Exp Mar Biol Ecol 73(3):225–242

Joo R, Dupré E (2002) Efecto de diferentes crioprotectores sobre la motilidad espermática de la macha *Mesodesma donacium* (Mollusca, Bivalvia). Investig Mar 30(2):75–79

Jovanovich MC, Marion KR (1989) Gametogenic cycle of *Rangia cuneata* (Mactridae, Mollusca) in Mobile Bay, Alabama, with comments on geographic variation. Bull Mar Sci 45(1):130–138

Junot JA, Poirrier MA, Soniat TM (1983) Effects of saltwater intrusion from the Inner Harbor Navigation Canal on the benthos of Lake Pontchartrain, Louisiana. Gulf Res Rep 7(3):247–254

Lepore ML, Penchaszadeh PE, Fernández Alfaya J, Herrmann M (2009) Aplicación de calceína para la estimación del crecimiento de la almeja amarilla Mesodesma mactroides Reeve, 1854. Rev Biol Mar Ocean 44(3):767–774

Katkansky SC, Warner RW, Poole RL (1969) On the occurrence of larval cestodes in the Washington clam, *Saxidomus nuttalli*, and the gaper clam, *Tresus nuttalli*, from Drakes Estero, California. Calif Fish Game 55(4):317–322

Katsanevakis S, Bogucarskis K, Gatto F, Vandekerkhove J, Deriu I, Cardoso AS (2012) Building the European Alien Species Information Network (EASIN): a novel approach for the exploration of distributed alien species data. BioInvasions Rec 1:235–245

Keen AM (1958) Sea shells of tropical West America. Marine mollusks from Baja California to Peru. Stanford University Press, Stanford, CA

Keen AM (1961) What is *Anatina*. Veliger 4:9–12

Keen AM (1963) Marine molluscan genera of Western North America. Stanford University Press, Stanford

Keen AM (1966) West American mollusk types at the British Museum (natural history), I: T. A. Conrad and the Nuttall collection. Veliger 8(3):167–172

Keen AM (1969) Superfamily Mactracea Lamarck, 1809. In: Moore RC (ed) Treatise on invertebrate paleontology. Geological Society of America and University of Kansas Press, Lawrence, KS, pp 595–610

Keen AM (1971) Seashells of tropical West American. Stanford University Press, Stanford, CA

Kellogg JL (1915) Ciliary mechanisms of lamellibranchs with descriptions of anatomy. J Morphol 26(4):625–701

Kershaw RC (1958) Tasmanian intertidal Mollusca. J Malacol Soc Aust 1:58–100

Kessley DW, Hitz CR (1971) Subtidal clam explorations in southeastern Alaska [abst.]. Proc Natl Shellfisheries Assoc 61:9

Kilburn RN (1971) On some species of the families Tonnidae, Hipponicidae, Buccinidae, Columbariidae, Fasciolariidae, Psammobiidae and Mactridae (Mollusca) in South African waters. Ann Natal Mus 20:48–497

Kilburn RN (1973) The type material of South African marine Mollusca in the Natal Museum collection. Part 1. Bivalvia. Ann Natal Mus 21:697–711

Kilburn RN, Rippey E (1982) Sea shells of southern Africa. South Africa, MacMillan Johannesburg

King PP (1832) Description of the Cirripedia, Conchifera and Mollusca in a collection formed by the officers of HMS Adventure and Beagle employed between the years 1826 and 1830 in surveying the southern coasts of South America, including the straits of Magalhaens and the coast of Tierra del Fuego. Zool J 5:332–349

Klompmaker AA (2009) Taphonomic bias on drill-hole predation intensities and paleoecology of Pliocene mollusks from Langenboom (mill), the Netherlands. PALAIOS 24(11):772–779

Kong L, Li Q (2009) Genetic evidence for the existence of cryptic species in an endangered clam *Coelomactra antiquata*. Mar Biol 156(7):1507–1515

Kostylev VE, Todd BJ, Fader GB, Courtney RC, Cameron GD, Pickrill RA (2001) Benthic habitat mapping on the Scotian Shelf based on multibeam bathymetry, surficial geology and sea floor photographs. Mar Ecol Prog Ser 219:121–137

Kowalewski M, Flessa KW (1995) Comparative taphonomy and faunal composition of shelly cheniers from northeastern Baja California, Mexico. Cienc Mar 21(2):155–177

Kowalewski M, Avila Serrano GE, Flessa KW, Goodfriend GA (2000) Dead delta's former productivity: two trillion shells at the mouth of the Colorado River. Geology 28(12):1059–1062

Krantic S, Dube F, Quirion R, Guerrier P (1991) Pharmacology of the serotonin-induced meiosis reinitiation in *Spisula solidissima* oocytes. Dev Biol 146(2):491–498

Krebs HJ (1864) The West-Indian marine shells, with some remarks: a manuscript printed for circulation between collectors. by [three asterisks printed instead of name.]. W. Laubswidow and C. Jørgensen, Nyokøbing, Falster.

Weinkauff HC (1841–1884) Die Gattung *Mactra*. In: Küster HC, Kobelt W, Weinkauff HC, eds. Systematisches Conchylien-Cabinet Von Martini Und Chemnitz 11(2), 124 pp, 36 pls. (pp 1–8, pls 2–4, 1841, Küster; pl 5, 1842, Küster; pp 1–36 [1–8 reissue], pls 1, 6, 1879, Weinkauff; pp 9–36, pls 7–12, 1880, Weinkauff; pp 37–68, pls 13–24, 1881, Weinkauff; pp 69–92, pls 25–30, 1882, Weinkauff; pp 93–124, pls 31–36, 1884, Weinkauff)

Laitano MV, Chiaradia NM, Nuñez JD (2019) Clam population dynamics as an indicator of beach urbanization impacts. Ecol Indic 101:926–932

Lamarck JBPAdMd (1799) Prodrome d'une nouvelle classification des coquilles, comprenant une rédaction appropriée des caractères génériques, et l'établissement d'un grand nombre de genres nouveaux. Mémoires Société d'Histoire Naturelle de Paris 1:63–91

Lamarck JBPAdM (1809) Philosophie Zoologique, ou Exposition des Considérations Relative à l'Histoire Naturelle des Animaux; à la Diversité de Leur Organisation et des Facultés qu'ils en Obtiennent; aux Causes Physiques qui Maintiennent en Eux la Vie et Donnent lieu aux Mouvements qui'ils Exécutent; enfin, á celles qui Produisent, les Unes le Sentiment, et les Autres l'Intelligence de Ceux qui eu sont Doués. Dentu [et] L'Auteur, Paris

Lamarck, JBPAdM (1818) Histoire Naturelle des Animaux sans Vertèbres, Présentant les Caractères Généraux et Particuliers de ces Animaux, leur Distribution, leurs Classes, leurs Familles, leurs Genres, et la Citation des principales Espèces qui s'y Rapportent; Précédée d'une Introduction Offrant la Détermination des Caractères Essentiels de l'Animal, sa Distinction du Végétal et des Autres Corps Naturels; enfin, l'Exposition des Principes Fondamentaux de la Zoologie. Verdière, Paris. First edition, Vol. 5

Lamarck JBPAdMd (1815–1822) Histoire Naturelle des animaux sans vértebrés. Verdière, Paris

Lamprell KL, Healy JM (1998) Bivalves of Australia. Backhuys, Leiden

Lamprell KL, Whitehead T, Healy JM (1992) Bivalves of Australia. Crawford House Press, Goolwa, SA

Lamy E (1912a) Sur les espèces de Lamarck appartenant au genre *Mesodesma* Deshayes. Bulletin du Museum National d'Histoire Naturelle 18(4):245–254

Lamy E (1912b) Note sur le *Mesodesma mactroides* Deshayes. Bulletin du Museum National d'Histoire Naturelle 18(5):312–316

Lamy E (1913) Notes sur les especes rangees par Lamarck dans son genre *Lutraria*. Bulletin du Muesum National d'Histoire Naturelle 19:343–349

Lamy E (1914a) Notes sur les especes de genre *Mactra* decrites par Lamarck. Bulletin du Museum National d'Histoire Naturelle 20:127–135/239–247

Lamy E (1914b) Révision des *Mesodesmatidae* vivants du Muséum d'Histoire Naturelle de Paris. J Conchyliol 62:1–74

Lamy E (1916) Les mactres et les lutraires de la Mer Rouge (d'après les matériaux recueillis par M. le Dr Jousseaume). Bulletin du Muséum National d'Histoire Naturelle 22(5–6):237–242

Lamy E (1917–1918) Révision des Mactridae vivants du Muséum d'Histoire Naturelle de Paris. J Conchyliol 63[(4)17](3):173–275 (1917); (4):291–411 (1918)

Lamy E (1925) Notes sur les especes rangees par Lamarck dans son genre *Anatina*. Bulletin du Muesum National d'Histoire Naturelle 31:372–378

Laurent L (1971) The spawning cycle and juvenile growth rates of the gaper clam, *Tresus nuttallii*, of Elkhorn Slough, California. Western Society of Malacologists. Echo 4:24–25

Lauzier RB, Hand CM, Campbell A, Heizer S (1998) A review of the biology and fisheries of the horse clams (*Tresus capax* and *Tresus nuttallii*). Canadian Stock Assessment Secretariat, Research Document N° 898

Lermond NW (1936) Checklist of Florida marine shells. Privately Published, Gulfport, FL

Lepore ML, Penchaszadeh PE, Fernández Alfaya J, Herrmann M (2009) Aplicación de calceína para la estimación del crecimiento de la almeja amarilla Mesodesma mactroides Reeve, 1854. Rev Biol Mar Ocean 44(3):767–774

Levinton J (1970) The paleoecological significance of opportunistic species. Lethaia 3:69–78

Li CC (1930) The Miocene and recent mollusca of Panama Bay. Bull Geol Soc China 9(3):249–296

Lightfoot J (1786) A Catalogue of the Portland Museum, Lately the Property of the Duchess Dowager of Portland, deceased: which will be sold by auction, by Mr. Skinner and Co. on Monday the 24th of April, 1786, and the thirty-seven following days, at twelve O'Clock, Sundays, and the 5th of June, (the day his Majesty's birth-day is kept) excepted; at her late dwelling house, in Privy-Garden, Whitehall; by order of the acting executrix. To be viewed ten days preceding the Sale. Catalogues may now be had on the premises, and of Mr. Skinner and Co., Aldersgate–Street, Price Five Shilling, which will admit the Bearer during the Time of the Exhibition and Sale. London

Lima M, Brazeiro A, Defeo O (2000) Population dynamics of the yellow clam *Mesodesma mactroides*: recruitment variability, density-dependence and stochastic processes. Mar Ecol Prog Ser 207:97–108

Lindell S, Walton B, Simmons J, Roberts S (2006) Hybridization between two genera of clams, *Spisula solidissima* and *Mulinia lateralis*, and earth growth trials. Abstract of Technical papers. J Shellfish Res 25(2):750

Linse K (1999) Mollusca of the Magellan region. A checklist of the species and their distribution. Sci Mar 63:399–407

Linnaeus CA (1758) Caroli Linnaei, Equitis de Stella Polari, Archiatri Regii, Med. and Botan. Profess. Upsal.; Acad. Upsal. Holmens. Petropol. Berol. Imper. Lond. Monspel. Tolos. Florent. Soc. Systema Naturae per Regna Tria Naturae, Secundum Classes, Ordines, Genera, Species, cum Characteribus, Differentiis, Synonymis, Locis. Editio Decima, reformata, 2 vols., paged continuously. Tomus I, Regnum Animale. [4] + p. 1-823 + [1 errata] p. Impensis Laurentii Salvii, Holmiae [Lipsiae, Stockholm]

Linnaeus CA (1767) Caroli a Linné, Equitis Aur. de Stella Polari, Archiatri Regii, Med. and Botan. Profess. Upsal., Acad. Paris. Upsal. Holmens. Petropol. Berol. Imper. Lond. Angl. Monspel. Tolos. Florent. Edinb. Bern. Soc. Systema Naturae per Regna Tria Naturae, Secundum Classes, Ordines, Genera, Species, cum Characteribus, Differentiis, Synonymis, Locis. Editio duodecima [12], reformata. Laurentii Salvii, Holmiae [Stockholm], 3 vols

Locard EAA (1890) Description des espèces française appartenant au genre *Mactra*. Bulletins Société Malacologique De France 7:1–76

Longo FJ, Anderson E (1969) Spermiogenesis in the surf clam Spisula solidissima with special reference to the formation of the acrosomal vesicle. J Ultrastruct Res 27(5–6):435–443

Loosanoff VL, Davis HC (1963) Rearing of bivalve mollusks. Adv Mar Biol 1:1–136

Lopes RP, Simone LRL, Dillenburg SR, Schultz CL, Pereira JC (2013) A Middle Pleistocene marine molluscan assemblage from the southern coastal plain of Rio Grande do Sul State, Brazil. Rev Bras Paleontol 16(3):343–360

López de Mesa LA, Cantera JR (2015) Marine mollusks of Bahía Málaga, Colombia (Tropical Eastern Pacific). Check List 11(1):1–18

López Z, Cárdenas L, González MT (2014) Metazoan symbionts of the yellow clam, *Mesodesma donacium* (Bivalvia), in Southern Chile: geographical variations. J Parasitol 100(6):797–805

Lucas M (1974a) The Mactridae of European coasts. La Conchiglia 6(67–68[9–10]):3–19

Lucas M (1974b) The Mactridae of European coasts. Part II. La Conchiglia 6(69–70[11–12]):16–20

Lutaenko KA (2014) Notes on the type material of *Mactra sulcataria* Deshayes in Reeve, 1854 (Bivalvia: Mactridae) and taxonomic history of the species. Bull Russ Far East Malacol Soc 18:97–106

MacGinitie GE (1941) On the method of feeding of four pelecypods. Biol Bull 80(1):18–25

Machell JR, DeMartini JD (1971) An annual reproductive cycle of the gaper clam, *Tresux capax* (Gould), in South Humboldt Bay, California. Calif Fish Game 57(4):274–282

Macpherson JH, Gabriel CJ (1962) Marine molluscs of Victoria. Melbourne University Press and National Museum of Victoria, Melbourne

Manger GE (1934) The geology of San Quintin Bay. Johns Hopkins Univ Stud Geol 11:273–304

Manta G, Barreiro M, Ortega L, Defeo O (2016) The effect of climate variability on the abundance of the sandy beach clam (*Mesodesma mactroides*) in the Southwestern Atlantic. J Coast Res 33(3):531–536

Marins LF, Levy JA (1999) High genetic distance between marine bivalves of the genus *Mesodesma* inhabiting the Atlantic and Pacific coasts of South America. Comp Biochem Physiol A Mol Integr Physiol 124(3):313–319

Marks JG (1951) Miocene stratigraphy and paleontology of Southwestern Ecuador. Bull Am Paleontol 33(139):354–358

Marsh ME, Hopkins G, Fisher FM, Sass RL (1976) Structure of the molluscan bivalve hinge ligament, a unique calcified elastic tissue. J Ultrastruct Res 54(3):445–450

Martinelli JC, Soto LP, González J, Rivadeneira MM (2017) Benthic communities under anthropogenic pressure show resilience across the Quaternary. R Soc Open Sci 4(9):170796. https://doi.org/10.1098/rsos.170796

Martínez and Saez, F. 1869. Moluscos del Viaje al Pacífico Verificado de 1862 a 1865 por una Comisión de Naturalistas Enviada por el Gobierno Español, Parte Segunda: Bivalvos Marinos, Madrid

Martynov AV (2002) The shell collection of J. H. Chemnitz in the Zoological Institute, St.-Petersburg. Ruthenica 12:1–18

Marwick J (1931) The tertiary Mollusca of the Gisbone district. N Z Geol Surv Paleontol Bull 13:1–177

Marwick J (1952) *Mactrona*, a new name for *Mactrula*, Marwick, preoccupied. Proc Malacol Soc Lond 29(2/3):83

Masello A, Defeo O (1986) Determinación de la longitud de primera madurez sexual en *Mesodesma mactroides* (Deshayes, 1854). Comun Soc Malacol Urug 6(51):387–392

Matsukuma A (2001) On the authorship of some Japanese mactrid species (Mollusca: Bivalvia). Chiribotan 21(1–2):5–9

Matthew GF (1874) Note sur les mollusques de la Formation Post-Pliocène de l'Acadie. Ann Soc Malacol Belgique 9:33–50

Maul GG, Avdalović N (1980) Nuclear envelope proteins from Spisula solidissima germinal vesicles. Exp Cell Res 130(1):229–240

Maury CJ (1917) Santo Domingo type sections and fossils. Part 1. Bull Am Paleontol 5(29):1–251

Maury CJ (1920) Recent Molluscs of the Gulf of Mexico and Pleistocene and Pliocene species from the gulf status. Bull Am Paleontol 8:35–147

Maury CJ (1925) A further contribution to the paleontology of Trinidad (Miocene horizons). Bull Am Paleontol 10:153–402

Maury CJ (1928) *Trinitasia*, a new Molluscan genus from South America. Science 67:318

Mayer C (1867) Mollusques familles des mactrides et des pholadomyides. In: Schabelitz L (ed) Catalogue systematique et descriptif des fossiles des terrains tertiares qui se trouvent au Musee federal de Zurich. Schabelitz, Zurich

Merrill AS, Ropes JW (1969) The general distribution of the surf clam and ocean quahog. Proc Nat Shellfish Assoc 59:40–45

Merrill AS, Davis JD, Emery KO (1978) Latitudinal and bathymetric ranges of living and fossil Mesodesma arctatum (Bivalvia) with notes on habits and habitat requirements. Nautilus 92(3):108–112

Metcalf AL (1980) Fossil *Rangia cuneata* (Mactridae) in Eddy County, New Mexico. Nautilus 94(1):2–3

Middendorff AV (1849) Beiträge zu einer Malacozoologia Rossica. II Aufzählung und Beschreibung der zur Meeresfauna Russlands gehörigen einschaler. Memoires de L'Academie Imperiale des Sciences de Siant-Petersbourg 6(2):517–610

Mikkelsen PM, Bieler R (2007) Seashells of Southern Florida—living marine mollusks of the Florida keys and adjacent regions: bivalves. Princeton University Press, Princeton, NJ

Miller W (1983) Stratigraphy of newly exposed quaternary sediments, eastern Orleans Parish, Louisiana. Tulane Stud Geol Paleontol 17(4):85–104

Möller MCP (1832) Übersicht des Herzogl., sonst Schmidtischen Conchylien-Cabinets, im Kunst- und Naturalien-Cabinet zu Gotha. Isis von Oken. 1832(2): 127–136

Moore EJ (2002) Family Mactridae. Tertiary marine pelecypods of California and Baja California, Chapter F. http://members.peak.org/~chintimp/Mactridae.htm

Moore RC (1969) Treatise on Invertebrate Paleontology. Part N, Mollusca 6, Bivalvia, Vol. 1. Geological Society of America and the University of Kansas Press, Lawrence, Kansas. xxxviii + p. 1–489

Montagu G (1808) Supplement to Testacea Britannica. With additional plates. White, London, and Woolmer, Exeter

Mörch OAL (1853) Catalogus Conchyliorum, quae Reliquit D. Alphonso d'Aguirra & Gadea, Comes de Yoldi: Regis Danie Cubiculariorum Princeps, Ordinis Dannebrogici in Prima Classe & Ordinis Caroli Tertii Eques. Ludovici Kleini, Hafniae (Copenhagen). Part 2, Acephala, Annulata Cirripedia, Echinodermata

Mörch OAL (1861) Beitrage zur Molluskenfauna Central Amerika's. Malakozool Bl 7(4):170–192

Morris PA (1951) A field guide to the shells of our Atlantic and Gulf coasts, Peterson field guide series. Houghton Mifflin Co., Boston

Morrison JPE (1959) Brackish water genera of Mactridae. Annu Rep Am Malacol Union 1958:26

Morrison JPE (1965) New brackish water mollusks from Louisiana. Proc Biol Soc Wash 78:217–224

Morton B (2010) Form and functional morphology of *Raetellops pulchella* (Bivalvia: Mactridae) an example of convergent evolution with anomalodesmatans. Invertebr Biol 129(3):241–251

Moore RC (1969) Treatise on invertebrate paleontology. Part N, Mollusca 6, Bivalvia, Vol. 1. Geological Society of America and the University of Kansas Press, Lawrence, Kansas. xxxviii + p. 1–489

Moss SM (1989) Effects of exogenous androgens on growth, biochemical composition, and reproduction of the coot clam, *Mulinia lateralis*. Pac Sci 43(2):200

Munroe DM, Powell EN, Mann R, Klinck JM, Hofmann EE (2013) Underestimation of primary productivity on continental shelves: evidence from maximum size of extant surfclam (Spisula solidissima) populations. Fish Oceanogr 22(3):220–233

Musmeci L, Hernández M, Scolaro J, Bala L (2013) Almejas en golfos norpatagónicos, Argentina: disponibilidad trófica para aves playeras migratorias. Rev Biol Mar Oceanogr 48(2):219–225

Nakazima M (1965) On the differentiation of the crenated-folds in the midgut-gland of Eulamellibranchia (VIII). Crenated-folds in some species of deep burrowers. Venus 24(1):58–66

Narchi W (1981) Aspects of the adaptive morphology of Mesodesma mactroides (Bivalvia: Mesodesmatidae). Malacologia 21(1–2):95–110

Narváez DA, Munroe DM, Hofmann EE, Klinck JM, Powell EN, Mann R, Curchitser E (2015) Long-term dynamics in Atlantic surfclam (Spisula solidissima) populations: the role of bottom water temperature. J Mar Syst 141:136–148

Navarro JM, Labarta U, Fernández-Reiriz MJ, Velasco A (2003) Feeding behavior and differential absorption of biochemical components by the infaunal bivalve *Mulinia edulis* and the epibenthic *Mytilus chilensis* in response to changes in food regimes. J Exp Mar Biol Ecol 287(1):13–35

Neumayr M (1884) Zur Morphologie des Bivalven Schlosses. Akademie der Wissenschaften, Wien (Mathematisch-Naturwissenschaftlichen Classe), Sitzungsberichte 88(1):385–419

Nevesskaja LA (2009) Principles of systematics and the system of bivalves. Paleontol J 43:1–11
Nevesskaja LA, Scarlato OA, Starobogatov YI, Eberzin AG (1971) New ideas on bivalve systematics. Paleontol J 5:141–155
Nicklès M (1950) Mollusques testacés marins de la côte occidentale d'Afrique, Manuels ouest-Africains. Lechevalier, Paris
Nielsen SN, Valdovinos C (2008) Early Pleistocene mollusks of the Tubul Formation, south-central Chile. Nautilus 122(4):201–216
Norse EA (1993) Global marine biological diversity: a strategy for building conservation into decision making. Island Press, Washington, DC
Nowell-Usticke GW (1969) A supplementary listing of new shells (illustrated). To be added to the Check List of the marine shells of St. Croix. Lane Press, Burlington, Vermont
Nowell-Usticke GW (1971) A supplementary listing of new species (illustrated), revised edition, to be added to the Checklist of the Marine Shells of St. Croix. Nowell-Usticke, Christiansted, St. Croix, U. S. Virgin Islands. Privately published
Oken L (1816) Okens Lehrbuch der Zoologie. In Lorenz Oken, 1815–1816, Okens Lehrbuch der Naturgeschichte, 2 vols
Okutani T, Masaru T, Horikawa H (1989) Bivalves from continental shelf and slope around Japan. Japanese Fisheries Resources Conservation Association, Tokyo
Oliva D, Abarca A, Gutiérrez R, Celis A, Herrera L, Pizarro V (2013) Effect of stocking density and diet on growth and survival of post-larvae of the taquilla clam *Mulinia edulis* cultivated in sand in a hatchery. Rev Biol Mar Oceanogr 48(1):37–44
Oliver PG (1992) Bivalved seashells of the Red Sea. Hemmen, Cardiff
Oliver PG, Morgenroth H, Salvador A (2017) Type specimens of Mollusca described by Col. George Montagu in the Royal Albert Memorial Museum and Art Gallery, Exeter and The Natural History Museum, London. Zoosyst Evol 93(2):363–412
Olivier S, Penchaszadeh PE (1968) Evaluación de los efectivos de almeja amarilla (*Mesodesma mactroides* Desh, 1854) en las costas de la provincia de Buenos Aires. Proy Des Pesq FAO Serv Inf Tecn 8:1–19
Olivier S, Capezzani D, Carreto J, Christiansen H, Moreno V, de Moreno JA, Penchaszadeh PE (1971) Estructura de la comunidad, dinámica de la población y biología de la almeja amarilla (*Mesodesma mactroides*) en Mar Azul. Proy Des Pesq FAO Serv Inf Tecn 27:1–90
Olsson AA (1922) The Miocene of northern Costa Rica. Bull Am Paleontol 9(39):1–309
Olsson AA (1932) Contributions to the tertiary paleontology of Northern Peru: Part 5, the Peruvian Miocene. Bull Am Paleontol 19(68):5–216
Olsson AA (1935) *Mactra alata var subalata* on the west coast. Nautilus 48(3):105
Olsson AA (1961) Mollusks of the tropical eastern Pacific - Panamic Pacific Pelecypoda. Ithaca, New York. 574 pp
Olsson AA (1964) Neogene mollusks from Northwestern Ecuador. Ithaca, New York. 256 pp
Olsson AA, Petit RE (1964) Some neogene mollusca from Florida and the Carolinas. Bull Am Paleontol 47(217):509–574
Olszewski TD, Kaufman DS (2015) Tracing burial history and sediment recycling in a shallow estuarine setting (Copano Bay, Texas) using postmortem ages of the bivalve *Mulinia lateralis*. PALAIOS 30:224–237
Ortiz MA, Stotz WB (1996) Distribución de juveniles de asentamiento reciente de *Mesodesma donacium* (Lamarck, 1818) (Mollusca: Bivalvia: Mesodesmatidae) en tres bahías de la IV Región: variables físicas, químicas y biológicas que le caracterizan. Biología Pesquera 25:27–40
Ortiz MA, Stotz WB (2003) Estimación preliminar del número óptimo de muestras para la evaluación del reclutamiento en *Mesodesma donacium* (Lamarck, 1818) en playas del centro-norte de Chile: aplicación del análisis de poder. Cienc Mar 29(3):371–375
Osorio C, Atria J, Mann S (1979) Moluscos Marinos de importancia económica en Chile. Biología Pesquera 11:3–47

Packard EL (1916) Mesozoic and Cenozoic Mactrinae of the Pacific coast of North America. Bull Depart Geol Univ Calif Publ 9(15):261–360

Paetel F (1883) Catalog der Conchylien Sammlung von Fr. Paetel, Berlin. Verlag Von Gebrüder Paetel

Palacio EF, Gracia A, Duque G (2010) Benthic Mollusks of La Guajira (10 and 50 m depths), Colombian Caribbean. Boletín de Investigaciones Marinas y Costeras-INVEMAR 39(2):397–416

Palazzo RE, Vaisberg E, Cole RW, Rieder CL (1992) Centriole duplication in lysates of *Spisula solidissima* oocytes. Science 256(5054):219–221

Palm EC, Esler D, Anderson EM, Wilson MT (2012) Geographic and temporal variation in diet of wintering White-winged Scoters. Waterbirds 35(4):577–590

Paramonova NP (1982) O zavisimosti stroeniia zamka maktrin (Bivalvia: Mactridae) ot solenosti basseinov [Relationship between hinge structure in Mactrinae (Bivalvia: Mactridae) and water salinity]. In: Kafanov A (ed) Biologiia Shel'fovykh zoi Mirovogo Okeana [Biology of the shelf zones of the world] 2. Vladivostok, Akademiia Nauk SSSR, Dalinevostochykh Nauchnykh Teentr, Institute Biologia Moria, Moria, pp 41–42

Paredes C, Cardoso F, Santamaría J, Esplana J, Llaja L (2016) Lista anotada de los bivalvos marinos del Perú. Rev Peru Biol 23(2):127–150

Parodiz JJ (1962) Los Moluscos marinos del Pleistoceno Rioplatense. Comun Soc Malacol Urug 1(2):29–48

Passos FD, Magalhães FT (2011) A comparative study of the Bivalvia (Mollusca) from the continental shelves of Antarctica and Brazil. Biota Neotrop 11(1):143–155

Pastorino G (1995) Moluscos costeros Recientes de Puerto Pirámide, Chubut, Argentina. Academia Nacional de Ciencias, Córdoba, pp 1–30

Patterson TL (1933) Comparative physiology of the gastric hunger mechanism. Ann N Y Acad Sci 34(21):55–284

Patton JS, Quinn JG (1973) Studies on the digestive lipase of the surf clam *Spisula solidissima*. Mar Biol 21:59–69

Pearce JB (1965) On the distribution of *Tresus nuttalli* and *Tresus capax* (Pelecypoda: Mactridae) in the waters of Puget Sound and the San Juan Archipelago. Veliger 7(3):166–170

Pearce JB (1966) On *Pinnixa faba* and *Pinnixa littoralis* (Decapoda: Pinnotheridae) symbiotic with the clam, *Tresus capax* (Pelecypoda: Mactridae). In: Barnes H (ed) Some contemporary studies in marine science. Hafner, New York, pp 565–589

Pekar SF, McHugh CM, Christie-Blick N, Jones M, Carbotte SM, Bell RE, Lynch-Stieglitz J (2004) Estuarine processes and their stratigraphic record: paleosalinity and sedimentation changes in the Hudson Estuary (North America). Mar Geol 209(1–4):113–129

Peredo S, Parada E, Valdebenito I (1987) Gametogenesis and reproductive cycle of the surf clam *Mesodesma donacium* (Lamarck, 1818) (Bivalvia: Mesodesmatidae) at Queule Beach, southern Chile. Veliger 30(1):55–68

Pérez EP, Chávez VJE (2004) Modelling short-term dynamic behavior of the surf clam (*Mesodesma donacium*) fishery in northern Chile using static and dynamic catchability hypotheses. Interciencia 29(4):196–198

Pérez J, Alfonsi C, Salazar S, Macsotay O, Barrios J, Martinez R (2007) Especies Marinas exóticas y criptogénicas en las costas de Venezuela. Boletin del Instituto de Oceanografía de Venezuela 46:79–96

Perrilliat MC (1992) Bivalvos y gasterópodos de la formacion Ferrotepec (Mioceno Medio) de Michoacan. Paleontologia Mexicana 60:1–48

Perry LM (1940) Marine shells of the Southwest Coast of Florida. Bull Am Paleontol 26(95):1–260

Peterson CH (1977) Competitive organization of the soft-bottom macrobenthic communities of southern California lagoons. Mar Biol 43(4):343–359

Peterson CH, Andre SV (1980) An experimental analysis of interspecific competition among marine filter feeders in a soft sediment environment. Ecology 61(1):129–139

Petit RE (2012) John Edward Gray (1800–1875): his malacological publications and molluscan taxa. Zootaxa 3214:1–125

Petit RE, Coan EV (2008) The molluscan taxa made available in the Griffith and Pidgeon (1833–1834) edition of Cuvier, with notes on the editions of Cuvier and on Wood's Index Testaceologicus. Malacologia 50:219–264

Petit de la Saussaye S (1853a) Description de deux nouvelles especes du genre *Gnathodon*. J Conchyliol 4:83–84

Petit de la Saussaye S (1853b) Description de deux especes appartenant au genre *Gnathodon* Gray; Rangia Ch. Desmoulins. J Conchyliol 4:163–168

Petuch EJ (1998) The molluscan fauna of the Wawa River Region, Miskito Coast, Nicaragua: ecology, biogeographical implications, and descriptions of new taxa. Nautilus 111(1):22–44

Petuch EJ, Roberts CE (2007) The geology of the Everglades and adjacent areas. CRC Press, Boca Raton, FL

Philippi RA (1844–1850) [Monograph of] *Mactra*. In: Abbildungen Und Beschreibungen Neuer Oder Wenig Gekannter Conchylien. T. Fischer, Cassel. (vol 1(7), pp 165–167, pl 1 (1844); vol 2(3), pp 71–74, pl 2 (1846); vol 3(8), pp 135–138, pl 3 (1850))

Philippi RA (1845) Diagnosen einiger neuen conchylien. Arch Natur 11:50–71

Philippi RA (1849) Centuria tertia testaceorum novorum. Z Malakozool 5(10):151–160

Philippi RA (1887) Die Tertiären und Quartären Versteinerungen Chiles. Brockhaus, Leipzig, 266 pp, 58 pls. [Mollusca: pp 33–216, pls 1–47]. Published simultaneously in Spanish as: 1887b, Los fósiles Terciarios i Cuatarios de Chile. Gobierno de Chile, Santiago, 256 pp. [Mollusca: pp 30–209, same pls]

Philippi RA (1893) Las especies chilenas del género Mactra. Anales del Museo Nacional de Chile Zoologia 1893:1–15

Pilsbry HA (1931) The Miocene and recent mollusca of Panama Bay. Proc Acad Nat Sci Phila 83(for 1931):427–440

Pilsbry HA, Johnson CW (1917) New mollusca of the Santo Domingan Oligocene. Proc Acad Natl Sci Phila 69(for 1917):150–202

Pilsbry HA, Lowe HN (1932) West Mexican and Central American mollusks collected by H. N. Lowe, 1929–31. Proc Acad Natl Sci Phila 84:33–144

Pilsbry HA, Olsson AA (1935) New mollusks from the Panamic Province. Nautilus 49(1):16–19

Pilsbry HA, Olsson AA (1941) A Pliocene fauna from western Ecuador. Proc Acad Natl Sci Phila 93:1–79

Pilsbry HA (1897) List of Mollusks collected in Maldonado Bay, Uruguay by Dr. Wm. H. Rush, U.S.N. The Nautilus 11:6 9

Pohlo RH (1964) Ontogenetic changes of form and mode of life in *Tresus nuttalli* (Bivalvia: Mactridae). Malacologia 1(3):321–330

Poorman FL, Poorman LH (1978) Additional molluscan records from Bahía de Los Angeles, Baja California Norte. Veliger 20(4):369–374

Portell RW, Schindler KS, Morgan GS (1992) The Pleistocene molluscan fauna from Leisley Shell Pit 1, Hillsborough County, Florida. Florida Geological Survey Special Publication N° 36: Plio-Pleistocene stratigraphy and paleontology of southern Florida

Porter HJ, Schwartz FJ (1981) Trawl collections of *Macoma constricta* and *Spisula raveneli* (Bivalvia: Tellinidae and Mactridae) in vicinity of Cape Fear River, NC and their relationship to periods of environmental stress. Bull Am Malacol Union 1981:15–19

Potiez VLV, Michaud ALG (1844) Galerie des Mollusques, ou Catalogue Méthodique, Descriptif et Raisonné des Mollusques et Coquilles du Muséum de Douai, vol 2. Baillière, Paris

Poutiers J (1995) Bivalves (Acephala, Lamellibranchia, Pelecypoda). In: Fishcer W, Krupp F, Schneider W, Sommer C, Carpenter KE, Niem VH (eds) FAO species identification guide for fishery purposes, The living marine resources of the Western Central Pacific, vol 1. Food and Agriculture Organization (FAO) of the United Nations, Rome, pp 99–222

Powell CL, Grant LB, Conkling SW (2005) Paleoecologic analysis and age of a late Pleistocene fossil assemblage from Upper Newport Bay, Newport Beach, Orange County, California. Veliger 47(3):183–192

Prezant RS, Counts CL III, Chapman EJ (2002) Mollusca of Assateague Island, Maryland and Virginia: additions to the fauna, range extensions, and gigantism. Veliger 45(4):337–355

Prime T (1860) Synonymy of the known species of *Rangia*, a genus of the family Mactracea. Proc Boston Soc Nat Hist 7:347–348
Prior DJ (1974) Role of the incurrent siphonal valve in the surf clam, *Spisula solidissima* (Mactridae). Nautilus 88(4):115–117
Pulteney R (1799) Catalogues of the birds, shells, and some of the more rare plants, of Dorsetshire. From the new and enlarged edition of Mr. Hutchins's History of that Country. Nichols, London
Pulteney R (1813) Catalogue of the Birds, Shells, and Some of the More Rare Plants, of Dorsetshire. From the New and Enlarged Edition of Mr. Hutchins's History of that County. With Additions [by William Rackett] and a Brief Memoir of the Author. Folio edition, Nichols, Son, and Bentley, London
Purchon RD (1960) The stomach in the Eulamelibranchia; stomach types IV and V. Proc Zool Soc Lond 135:431–489
Purchon RD (1987) Classification and evolution of the Bivalvia: an analytical study. Philos Trans R Soc Lond B Biol Sci 316:277–302
Qi Z (2004) Mactridae. In: Qi Z (ed) Seashells of China. China Ocean Press, Beijing, China, pp 268–273, pls 148–150
Rafinesque C (1814) Definizioni cli 36 nuovi generi di animali marini della Sicilia. Specchio delle scienze, o Giornale enciclopedico di Sicilia. Palermo
Ramírez R, Paredes C, Arenas J (2003) Moluscos del Perú. Rev Biol Trop 51(3):225–284
Rang M (1834) Mémoire sur Le genre Gnatodon et description de son animal. Nouvelle Ann Mus Hist Nat 3(3):217–234
Rapoport EH (1960) Algunas observaciones sobre la almeja *Mesodesma mactroides* de Monte Hermoso (Argent.). Primer Congreso Sudamericano de Zoología 2:235–243
Ravenel E (1834) Catalogue of the recent shells in the cabinet of Edmund Ravenel. Charleston, SC
Reagan AB (1909) Some notes on the Olympic Peninsula, Washington. Trans Kansas Acad Sci 22:131–238
Rebhun LI (1962) Electron microscope studies on the vitelline membrane of the surf clam, *Spisula solidissima*. J Ultrastruct Res 6(1):107–122
Récluz MC (1852) Description de coquilles nouvelles. J Conchiliol 3:249–256
Reeve LA (1854a) Monograph of the genus *Mactra*. In: Reeve LA (ed) Conchologia iconica; or, Illustrations of the shells of molluscous animals. L. A. Reeve, London. [unnumbered pages are captions for the plates]. p 8, 21 pls
Reeve LA (1854b) Monograph of the genus *Lutraria*. In: Reeve LA (ed) Conchologia iconica; or, Illustrations of the shells of molluscous animals. L. A. Reeve, London. [unnumbered pages are captions for the plates]. p 8, 5 pls
Reeve LA (1854c) Monograph of the genus *Mesodesma*. In: Reeve LA (ed) Conchologia iconica; or, Illustrations of the shells of molluscous animals. L. A. Reeve, London. [unnumbered pages are captions for the plates]. p 8, 4 pls
Reeve LA (1873) Monographs of the genus *Gnathodon*. In: Reeve LA (ed) Conchologia iconica: or, Illustrations of the shells of moluscous animals. L. A. Reeve, London. [unnumbered pages are captions for the plates] p 19, 1 pl
Reguero M, Raz-Guzmán A (2018) Molluscs (Mollusca: Gastropoda, Bivalvia, Polyplacophora) of Laguna Madre, Tamaulipas, Mexico: spatial and temporal distribution. Gulf Mex Sci 34(1):32–55
Reid RG (1969) Seasonal observations on diet, and stored glycogen and lipids in the horse clam, *Tresus capax* (Gould, 1850). Veliger 11(4):378–381
Reid DG, Osorio C (2000) The shallow-water marine mollusca of the Estero Elefantes and Laguna San Rafael, southern Chile. Bull Nat Hist Mus Lond Zool 66:109–146
Reid RG, Rauchert K (1976) Catheptic endopeptidases and protein digestion in the horse clam *Tresus capax* (Gould). Comp Biochem Physiol B Comp Biochem 54(4):467–472
Reunov AA, Lutaenko KA, Zakhova EE, Vekhova EE, Reunova YA, Alexandrova YN, Sharina SN, Adrianov AV (2014) Disproportional heteromorphism of male gametes in the bivalve mollusk *Mactra chinensis* is related to genetic divergence of this species. Doklady Biol Sci 455:132–135

Riascos JM, Guzman N, Laudien J, Heilmayer O, Oliva M (2007) Suitability of three stains to mark shells of *Concholepas concholepas* (Gastropoda) and *Mesodesma donacium* (Bivalvia). J Shellfish Res 26(1):43–49

Riascos JM, Carstensen D, Laudien J, Arntz WE, Oliva ME, Güntner A, Heilmayer O (2009) Thriving and declining: climate variability shaping life-history and population persistence of *Mesodesma donacium* in the Humboldt Upwelling System. Mar Ecol Prog Ser 385:151–163

Richards HG (1962a) The cretaceous fossils of New Jersey. N J Paleontol Ser 61(2):1–237

Richards HG (1962b) Studies on the marine Pleistocene, Part 1. The marine Pleistocene of America and Europe. Part 2. The marine Pleistocene molluscs of Eastern North America. Am Philos Trans 52(3):1–141

Richards HG, Craig JR (1963) Pleistocene Mollusks from the continental shelf off Argentina. Proc Acad Natl Sci Phila 115:127–152

Rios EC (1966) Provisional list of Rio Grande do Sul Marine Mollusks. Notas e Estudos Escola de Geologia 1:15–40

Rios EC (1969) Additions and corrections to the provisional list of Rio Grande do Sul Marine Mollusks. Rio Grande, Rio Grande do Sul

Rios EC (1970) Coastal Brazilian seashells. Museu Oceanografico de Rio Grande, Rio Grande

Rios EC (1975) Brazilian marine mollusks iconography. Fundação Universidade do Rio Grand, Rio Grande

Rios EC (1985) Seashells of Brazil Rio Grande. Fundação Universidade do Rio Grande, Rio Grande

Rios EC (1994) Seashells of Brazil, 2nd edn. Editora da Fundaçao Universidade do Rio Grande, Rio Grande

Rios EC (2009) Compendium of Brazilian sea shells. Rio Grande, Rio Grande do Sul, Rio Grande

Robinson AM, Breese WP (1982) The spawning season of four species of clams in Oregon. J Shellfish Res 2(1):55–57

Robinson DG, Montoya M (1987) Los moluscos marinos de la costa Atlántica de Costa Rica. Rev Biol Trop 35(2):375–400

Robinson WE, Wehling WE, Morse MP (1984) The effect of suspended clay on feeding and digestive efficiency of the surf clam, *Spisula solidissima* (Dillwyn). J Exp Mar Biol Ecol 74(1):1–12

Rochebrune AT, Mabille J (1889) Mollusques. In: Ministères de la Marine et de li Instruction Publique, ed. Mission Scientifique du Cap Horn 1882–1883. Gauthiers-Villars, Paris. 6(Zoologic 2, part 8):1–129

Röding PF (1798) Museum Boltenianum sive Catalogus Cimeliorum e Tribus Regnis Naturae quae olim Collegerat Joa. [Joachim] Fried. Bolten... Pars Secunda Continens Conchylia sive Testacea Univalvia, Bivalvia, and Multivalvia. Johan Christi. Trappii, Hamburgi

Rodriguez CA, Flessa KW, Dettman DL (2001) Effects of upstream diversion of Colorado River water on the estuarine bivalve mollusc *Mulinia coloradoensis*. Conserv Biol 15(1):249–258

Rogers P, García-Cubas AS, Gutiérrez (1981) Evolutión gonádica a nivel histológico de *Rangia cuneata* (Gray, 1831) de la Laguna Pom, Campeche, México (Mollusca: Bivalvia). Anales Del Centro de Ciencias Del Mar Y Limnología 8(1):1–20

Römer E (1862) Kritische Uebersicht sämmtlicher Arten aus der Cythereengruppe Lioconcha Mörch. Malakozool Bl 9(2):137–150

Rooij-Schuiling LAd (1972) Systematic notes on the Mesodesmatidae (Mollusca, Bivalvia), and descriptions of a new species and a new subspecies. Zool Meded 46:55–68

Ropes JW (1968a) Reproductive cycle of the Surf Clam, *Spisula solidissima* in offshore of New Jersey. Biol Bull 135(2):349–365

Ropes JW (1968b) Hermaphroditism in the surf clam, *Spisula solidissima*. Proc Natl Shellfisheries Assoc 58:63–65

Ropes JW, Merrill AS (1966) The burrowing activities of the surf clam. Underw Nat 3(4):11–17

Ropes JW, Merrill AS (1973) To what extent do surf clams move? Nautilus 87(1):19–21

Ropes JW, O'Brien L (1980) A unique method of aging surf clams. Bull Am Malacol Union 1979:58–61

Ropes JW, Chamberlin JL, Merrill AS (1969) Surf clam fishery. In: Firth FE (ed) Encyclopedia of marine resources. Van Nostrand, Reinhold and Co., New York, pp 119–126

Rosenthal ET, Hunt T, Ruderman JV (1980) Selective translation of mRNA controls the pattern of protein synthesis during early development of the surf clam, *Spisula solidissima*. Cell 20(2):487–494

Roth B (1979) Late Cenozoic marine invertebrates from northwest California and southwest Oregon. Unpublished Ph.D. Doctoral dissertation, University of California, Berkeley

Roy K, Jablonski D, Valentine JW (1994) Eastern Pacific molluscan provinces and latitudinal diversity gradient: no evidence for Rapoport's rule. Proc Natl Acad Sci U S A 91:8871–8874

Ruiz JT, Cipirán AI, Oyola MC, Atarama JS (2012) Recursos de la zona béntica de la bahía de Sechura. Cienc Desarro 15(1):19–50

Russell-Hunter WD, Grant DC (1962) Mechanics of the ligament in the bivalve *Spisula solidissima* in relation to mode of life. Biol Bull 122(3):369–379

Salvador LB, Domaneschi O, Amaral ACZ, Morgado EH, Henriques SA (1998) Intertidal malacofauna of sandy beaches of São Sebastião Island (São Paulo, Brazil). Rev Bras Zool 15(4):1013–1035

Saul LR (1973) Evidence for the origin of the Mactridae (Bivalvia) in the Cretaceous. Univ Calif Publ Geol Sci 97:1–59

Saul LR (1991) Northeast Pacific Mactridae comprise six supraspecific taxa, none of them *Mactra* and *Spisula* [abst.]. Annu Rep West Soc Malacol 23:17

Savage NB (1976) Burrowing activity in *Mercenaria* (L.) and *Spisula solidissima* (Dillwyn) as a function of temperature and dissolved oxygen. Mar Behav Physiol 3(4):221–234

Savenkoff C et al (2007) Effects of fishing and predation in a heavily exploited ecosystem: comparing periods before and after the collapse of groundfish in the southern Gulf of St. Lawrence (Canada). Ecol Model 204(1–2):115–128

Say T (1822) An account of the marine shells of the United States. J Acad Nat Sci Phila 2(2):221–224, 225–248, 257–276, 302–320, 321–325

Scarabino V (1977) Moluscos del Golfo San Matias (Provincia de Río Negro, República Argentina). Inventario y Claves para su identificacion. Comun Soc Malacol Urug 4(31–32):177–285

Scarabino F (2003) Lista sistemática de los bivalvia marinos y estuarinos de Uruguay. Comun Soc Malacol Urug 8(80–81):229–259

Scarabino F, Zaffaroni JC, Clavijo C, Carranza A, Nin M (2006) Bivalvos marinos y estuarinos de la costa Uruguaya: faunística, distribución, taxonomía y conservación. In: Menafra R, Rodríguez-Gallego L, Scarabino F, Conde D (eds) Bases para la conservación y el manejo de la costa uruguaya. Vida Silvestre Uruguay, Montevideo, pp 157–169

Scarabino F, Zelaya DG, Orensanz JM, Ortega L, Defeo O et al (2016) Cold, warm, temperate and brackish: bivalve biodiversity in a complex oceanographic scenario (Uruguay, southwestern Atlantic). Am Malacol Bull 33:284–301

Scarlato OA (1981) Dvustvorchatye molliuski umerennykh shirot zapadnoi chasti Tikhogo Okeana [Bivalve mollusks of the temperate latitudes of the western part of the Pacific Ocean]. Akademiia Nauk SSSR, Zoologicheskii Institut, Opredeliteli po Faune SSSR

Schrenk LI (1867) Mollusken des Amur-Landes und des Nordjapanischen Meeres. Reisen und Forschungen im Amur-Lande in den Jahren 1854–1856 im Auftrage der Kaiserl. Akad. Wissensch. zu St. Petersburg ausgeführt und in Verbindung mit mehreren Gelehrten herausgegeben. 2[Zool.] (3):259–974

Schuetz AW (1975) Induction of nuclear breakdown and meiosis in *Spisula solidissima* oocytes by calcium ionophore. J Exp Zool 191(3):433–440

Schumacher CF (1817) Essai d'un Nouveau Systéme des Habitations des Vers Testacés avec XXII Planches. L'Imprimerie de Ma. le directeur Schultz, Copenhagen, [3] + 20 p of text, + Table systématique des monothalmes, pp 21–30 [no genera therein], + Table abrégée des genres, pp 31–263, + p 264 (blank), + Table de noms François des genres et des espèces de coquilles, pp 265–271, + Index nominum Latinorum generum atque specierum testaceorum, pp 272–278, + Table des synonymes, pp 279–284, + Explication des figures, pp 285–286, + Errata, p 287, + 22 pls [in French with parallel French and Latin texts.]

Segura Zamudio M, Galindo Flores O, Flores D (1998) Evaluación del recurso macha (*Mesodesma donacium*) en el litoral de Ica y Arequipa, Marzo–Abril 1998. IMARPE, Informe Progresivo; N° 95

Sherborn CD (1940) Where is the collection? An account of the various natural history collections which have come under the notice of the compiler Charles Davies Sherborn between 1880 and 1939. Cambridge University Press, Cambridge

Shuttleworth RJ (1856) Description de nouvelles espéces, Premiere décade, espéces nouvelles pour la faune des Antilles. J Conchiliol 5(1):168–175

Signorelli JH (2012) The Molluscan Genera *Mactrella* and *Mactrinula*: taxonomic revision and redescription of type species. Malacologia 55(2):191–202

Signorelli JH, Carter JG (2016) The Anatinellidae and Kymatoxinae: a reassessment of their affinities within the superfamily Mactroidea (Mollusca, Bivalvia). Am Malacol Bull 33:204–211

Signorelli JH, Pastorino G (2006) Anatomy and functional morphology of the southwestern Atlantic bivalve *Mactra isabelleana* d'Orbigny, 1846 (Heterodonta: Mactridae) [abst.]. In: Malchus N, Pons JM (eds) Bivalvia 2006. International Congress on Bivalvia, Barcelona, Spain, July 22–27, 2006, Scientific program and abstracts, p 73. 87 pp

Signorelli JH, Pastorino G (2011) Revision of the Magellanic Mactridae Lamarck, 1809 (Bivalvia: Heterodonta). Zootaxa 2757:47–67

Signorelli JH, Pastorino G (2012a) A revision of the living Mactridae (Bivalvia: Autobranchia) from Northern Argentina and Uruguay. Am Malacol Bull 30(1):85–101

Signorelli JH, Pastorino G (2012b) Taxonomic revision of Brazilian Mactridae Lamarck, 1809 (Bivalvia: Cardiida). Zootaxa 3245:30–53

Signorelli JH, Scarabino F (2010) *Mactra guidoi* n. sp. and Mactra patagonica (Bivalvia: Mactridae) two long misunderstood species from south western Atlantic Ocean. Malacologia 52(1):31–42

Signorelli JH, Márquez F, Pastorino G (2013) Phenotypic variation of southwestern Atlantic clam Mactra isabelleana (Bivalvia: Mactridae). J Mar Biol Assoc United kingdom 93:511–517

Sikora WB, Sikora JP (1982) Ecological characterization of the benthic community of Lake Pontchartrain, Louisiana. Publication N° LSU-CEL-82-05. Louisiana State University, Baton Rouge, LA

Sikora WB, Sikora JP, Prior AM (1981) Environmental effects of hydraulic dredging for clam shells in Lake Pontchartrain, Louisiana. Publication N° LSU-CEL-81-18. Louisiana State University, Baton Rouge, LA

Simkins LM, Simms AR, Cruse AM, Troiani T, Atekwana EA, Puckette J, Yokoyama Y (2012) Correlation of early and mid-Holocene events using magnetic susceptibility in estuarine cores from bays along the northwestern Gulf of Mexico. Palaeogeogr Palaeoclimatol Palaeoecol 346–347:95–107

Skoglund CC (2000) *Raeta plicatella* (Lamarck, 1818). (Bivalvia: Mactridae) a first report of the Panamic province. Festivus 32(11):149–150

Smith EA (1881) Account of the zoological collections made during the survey of H.M.S. "Alert" in the Straits of Magellan and on the Coast of Patagonia. IV. Mollusca and molluscoidea. Proc Zool Soc London 1881:22–44

Smith EA (1885) Report on the Lamellibranchiata collected by H. M. S. Challenger during the years 1873–1876. Report on the Scientific Results of the Voyage of H.M.S. Challenger during the years 1873–76, Zoology, pp 1–341

Smith EA (1905) On a small collection of Mollusca from Tierra del Fuego. Proc Malacol Soc Lond 6(6):333–339

Smith EA (1910) On South African marine Mollusca, with descriptions of new species. Ann South Afr Mus 2:175–220

Smith EA (1914) A list of Australian Mactridae, with a description of a new species. Proc Malacol Soc Lond 11(2):137–151

Smith EA (1915) Mollusca Part I-Gastropoda Prosobranchia, Scaphopoda and Pelecypoda. British Antarctic (Terra Nova) expedition, 1910. Natural history report. British Museum, London, pp 61–112

Smith M (1937) Further notes upon tertiary and recent mollusks from Florida together with descriptions of new species. Nautilus 51(2):65–68

Smith LS, Davis JC (1965) Haemodynamics in *Tresus nuttallii* and certain other bivalves. J Exp Biol 43(1):171–180

Smith JA, Dietl GP (2019) Molluscan metacommunity dynamics in the Colorado River estuary, Mexico before upstream water diversion. Anthropocene 25:100194. https://doi.org/10.1016/j.ancene.2019.100194

Smith BC, Wikfors GH (1992) Phytoplankton pigments accumulated by the Arctic surfclam, *Macromeris polynyma* (Stimpson, 1860). J Shellfish Res 11(2):479–483

Smith JA, Handley JC, Dietl GP (2018) Effects of dams on downstream molluscan predator–prey interactions in the Colorado River estuary. Proc R Soc Boston B 285. https://doi.org/10.1098/rspb.2018.0724

Smriglio C, Mariottini P, Ciommei C (1994) Molluschi del Mar Tirreno centrale, contributo IX: segnalazione di un consistente spiaggiamento di *Lutraria* (Linné, 1758). Boll Malacol 30(5–6):189–194

Soot-Ryen T (1959) Reports of the Lund University Chile expedition 1948–49. Pelecypoda. Lund University Arsskrift 55(6):1–86

Sowerby J (1817) In: Sowerby J, Sowerby JdC (1812–1846). The mineral conchology of Great Britain; or Coloured figures and description of those remains of Testaceous animals or shells which have been preserved at various times and depths in the earth. 7 vols. B. Meredith, London. Published in parts. James Sowerby published numbers 1-66, 1812 to 1822; his son, James de Carle Sowerby, completed numbers 67-113, 1823–1846

Sowerby GB I (1821–1834) The genera of recent and fossil shells, for the use of students, in conchology and geology. London. Published in 42 numbers vol 1: pls 1-126 (1821-1825); vol. 2, pls 127-262 + text (unpaginated) (1825-1834)

Sowerby GB I (1825) A catalogue of the shells contained in the collection of the late Earl of Tankerville arranged according to the Lamarckian conchological system. London

Sowerby GB II (1846) Descriptions of tertiary fossil shells from South America. In: Smith EC (ed) C. Darwin's geological observations on South America. Smith, Elder, London, pp 249–264

Sowerby GB III (1892) Marine shells of South Africa: a catalogue of all the known species, with references to figures in various works, descriptions of new species and figures of such as are new, little known, or hitherto unfigured. Sowerby, London

Sowerby GB III (1897) Appendix to marine shells of South Africa. A catalogue of all the known species with references to figures in various works, descriptions of new species, and figures of such as are new, little known, or hitherto unfigured. Sowerby, London

Soyer J, Soyer-Gobillard MO, Thiriot-Quiévreux C, Bouvy M, Cahet G (1987) Chemoautrotrophic bacterial endosymbiosis in Spisula subtruncata (Bivalvia, Mactridae). Ultrastructure, metabolic significance and evolutionary implications. Symbiosis 3(3):301–314

Spasskii AA, Spasskaya LP (1954) Systematics of Hymenolepididae parasiting birds. Trudy Gel'mintologicheskoy Laboratorii Akademii Nauk SSSR 7:55–119

Spencer RS, Campbell LD (1987) The fauna and paleoecology of the Late Pleistocene marine sediments of southeastern Virginia. Bull Am Paleontol 92(327):1–124

Spengler L (1802) Beskrivelse over det toskallede conchylie-slaegt Mactra. Skrivter af Naturhistorie - Selskabet 5:92–128

Spieker EM (1922) The paleontology of the Zorritos formation of the north Peruvian oil fields. Johns Hopkins Univ Stud Geol 3:1–197

Staff GM, Powell EN (1999) Onshore–offshore trends in community structural attributes: death assemblages from the shallow continental shelf of Texas. Cont Shelf Res 19(6):717–756

Staff GM, Stanton RJ, Powell EN, Cummins H (1986) Time-averaging, taphonomy, and their impact on paleocommunity reconstruction: death assemblages in Texas bays. GSA Bull 97(4):428–443

Stasek CR (1963) Synopsis and discussion of the association of ctenidia and labial palps in the bivalved Mollusca. Veliger 6:91–97

Stearns REC (1891) Edible shell notes. Nautilus 5(3):25–26

Stearns REC (1894) The shells of the Tres Marias and other localities along the shores of Lower California and the Gulf of California. Proc U S Natl Mus 17(996):139–204

Stecher HA III, Krantz DE, Lord CJ III, Luther GW III, Bock KW (1996) Profiles of strontium and barium in *Mercenaria mercenaria* and *Spisula solidissima* shells. Geochim Cosmochim Acta 60(18):3445–3456

Steele DH, Green JM, Carter JG (1979) A biological and oceanographic study of the Atlantic southwest coast marine region. Report to Canadian Parks Service. Dept Indian and Northern Affairs, Ottawa, ON

Stempell W (1899) Die Muscheln der Sammlung Plate. Fauna Chilensis 2(1):217–250. Abhandlungen zur Kenntniss der Zoologie Chiles nach den Sammlungen von L. Plate

Stenzel HB, Krause EK (1957) In: Stenzel HB, Krause EK, Twining JT (1957) Pelecypoda from the type locality of the Stone City Beds (middle Eocene) of Texas. Pub Univ Texas 5704:1–237

Stewart RB (1930) Gabb's California Cretaceous and Tertiary type lamellibranchs. Acad Nat Sci Phila Spec Publ 3:1–314

Stimpson W (1860) Check lists of the shells of North America. 3. East coast: Arctic seas to Georgia. Smithsonian Miscellaneous Collections 2(128):1–6

Stoliczka F (1870, 1871) The Pelecypoda, with a review of all known genera of this class, fossil and Recent. In Thomas Oldham, Cretaceous Fauna of Southern India. Vol. 3, Palaeontologia Indica, Being Figures and Descriptions of the Organic Remains Procured During the Progress of the Geological Survey of India. Trübner and Co. Memoirs of the Geological Survey of India, Calcutta (series 1), no. 3, issued in 13 parts

Stout WE (1970) Some associates of *Tresus nuttallii* (Conrad, 1837) (Pelecypoda: Mactridae). Veliger 13(1):67–70

Strathmann MF (1987) Phylum Mollusca, Class Bivalvia. [chapter advisers: A.R. Kabat and D.Ó Foighil]. In: Strathmann MF (ed) Reproduction and development of marine invertebrates of the Northern Pacific Coast. University of Washington, Seattle, pp 309–353

Strong AM (1925) Notes on the mactras of the West Coast. Nautilus 38(3):98–102

Stump TE (1975) Pleistocene molluscan paleoecology and community structure of the Puerto Libertad region, Sonora, Mexico. Palaeogeogr Palaeoclimatol Palaeoecol 17(3):177–226

Sundberg K, Kennedy VS (1992) Growth and development in the Atlantic *Rangia*, *Rangia cuneata*. J Shellfish Res 11(1):9–12

Sundberg K, Kennedy VS (1993) Larval settlement of the Atlantic *Rangia*, *Rangia cuneata* (Bivalvia: Mactridae). Estuaries 16:223–228

Swan EF, Finucane JF (1952) Observations of the genus *Schizothaerus*. Nautilus 66(1):21–26

Swainson WJ (1840) A Treatise on Malacology; or the Natural Classification of Shells and Shell-fish. [Vol. 122 of "The Cabinet Cyclopaedia" edited by Dionysius Lardner.] A. Longman, Orme, Brown, Green and Longmans, and John Taylor, London

Swingle HA, Bland DG (1974) Distribution of the estuarine clam *Rangia cuneata* Gray in coastal waters of Alabama. Alabama Mar Resour Bull 10:9–16

Tate R (1889) Description of some new species of South Australia marine and freshwater Mollusca. Trans Proc Reprt Roy Soc S Aust 9:62–75

Tallarico LdF, Dias Passos F, Marcondes Machado F, Campos A, Recco-Pimentel SM, Orlandi Introíni G (2014) Bivalves of the São Sebastião Channel, north coast of the São Paulo State, Brazil. Check List 10(1):97–105

Taylor RL, Smith AC (1966) Polypoid and papillary lesions in the foot of the gaper clam, *Tresus nuttalli*. J Invertebr Pathol 8(2):264–266

Taylor JD, Kennedy WJ, Hall A (1973) The shell structure and mineralogy of the Bivalvia. II. Lucinacea-Clavagellacea conclusions. Bull Br Mus Nat Hist Zool 22:253–294

Taylor JD, Williams ST, Glover EA, Dyak O (2007) A molecular phylogeny of heterodont bivalves (Mollusca: Bivalvia: Heterodonta): new analyses of 18S and 28S rRNA genes. Zool Scr 36:587–606

Tenore KR, Horton DB, Duke TW (1968) Effects of bottom substrate on the brackish water bivalve *Rangia cuneata*. Chesap Sci 9(4):238–248

Thiele KHJ (1929, 1931, 1934, 1935) Handbuch der Systematischen Weichtierkunde. Gustav Fischer, Jena. Published in 4 parts, 1154 + vi p., 773 text-figs

Thompson GA, Sánchez De Bock MF (2009) Influence of beach morphodynamics on the bivalve *Donax hanleyanus* and *Mesodesma mactroides* populations in Argentina. Mar Ecol 30(2):198–211

Tomicic KJ (1978) *Raeta undulata* (Gould, 1851) una especie nueva para aguas Chilenas (Mollusca, Pelecypoda, Mactridae). Museo Nacional de Historia Natural de Chile Noticiario Mensual 22(257–258):11–12

Tomlin JRB (1921) Six new marine shells from South Africa. J Conchol 16(7):215-–217

Tomlin JRlB (1924) Notes on some Mactridae. J Conchol 17(5):134–136

Troiani BT, Simms AR, Dellapenna T, Piper E, Yokoyama Y (2011) The importance of sea-level and climate change, including changing wind energy, on the evolution of a coastal estuary: Copano Bay, Texas. Mar Geol 280(1–4):1–19

Tryon GW (1870) Description of new species of marine bivalve Mollusca in the collection of the Academy of Natural Sciences. Am J Conchol 5(3):170–172

Tryon GW (1873–1874) American Marine Conchology: or, descriptions of shells of the Atlantic Coast of the United States from Maine to Florida. Tryon GW, Philadelphia

Tsuchida E, Kurozumi T (1995) Fauna of marine mollusks of the sea around Otsuchi Bay, Iwati prefecture (5) Bivalvia, 2. Otsuchi Mar Res Cent Rep 20:13–42

Tuomey M, Holmes FS (1857) Pleiocene fossils of South-Carolina: containing descriptions and figures of the Polyparia, Echinodermata and Mollusca. Russell & Jones, Charleston, SC

Turgeon DD, Lyons WG, Mikkelsen P, Rosenberg G, Moretzsohn F (2009) Bivalvia (Mollusca) of the Gulf of Mexico. In: Felder DL, Camp DK (eds) Gulf of Mexico–origins, waters, and biota, Biodiversity. Texas A&M University Press, College Station, TX, pp 711–744

Turton W (1822) Conchylia Insularum Britannicarum. The shells of the British Islands systematically arranged. M. A. Nattali, Covent-Garden, and Combe and Son, Leicester

Urk RMv (1959) De *Spisula's* van het Nederlandse strand. Basteria 23(1–2):1–32

Urk RMv (1964) The British species of *Spisula*. J Conchol 25(7):268–275

Urk RMv (1980) Fossil and recent species of *Lutraria* (Mollusca, Bivalvia) in Europe, with descriptions of four new species. Mededelinger Van Der Werkgroep Voor Tertiaire En Kwartaire Geologie 17(4):235–266

Valentine JW (1957) Late Pleistocene faunas from the northwestern coast of Baja California, Mexico. Trans San Diego Soc Nat Hist 12(16):291–308

Vanatta EG (1903) A list of shells collected in western Florida and horn Island, Mississippi. Proc Acad Natl Sci Phila 55:756–757

Vázquez N, Cremonte F (2017) Review of parasites and pathologies of the main Bivalve Species of commercial interest of Argentina and Uruguay, Southwestern Atlantic Coast. Arch Parasitol 1(2):1–12

Vázquez N, Fiori S, Arzul I, Carcedo C, Cremonte F (2016) Mass mortalities affecting populations of the yellow clam *Amarilladesma mactroides* along its geographic range. J Shellfish Res 35(4):739–746

Vecchione M, Griffis RB (1996) How many species of surf clams? Oceanography 9(1):48–49

Velasco LA, Navarro JM (2002) Feeding physiology of infaunal (*Mulinia edulis*) and epifaunal (*Mytilus chilensis*) bivalves under a wide range of concentrations and qualities of seston. Mar Ecol Prog Ser 240:143–155

Verween A, Kerckhof F, Vincx M, Degraer S (2006) First European record of the invasive brackish water clam *Rangia cuneata* (G.B. Sowerby I, 1831) (Mollusca: Bivalvia). Aquat Invasions 1(4):198–203

Vivanco G, Oliva D, Abarca A (2014) Efecto de dietas en base a microalgas tradicionales, nativas y dietas artificiales sobre el crecimiento y supervivencia en larvas velígeras de la almeja taquilla, *Mulinia edulis*. Rev Biol Mar Oceanogr 49(2):339–349

Vogel JM, Stearns T, Rieder CL, Palazzo RE (1997) Centrosomes isolated from Spisula solidissima oocytes contain rings and an unusual stoichiometric ratio of α/β tubulin. J Cell Biol 137(1):193–202

Vokes HE (1968) Genera of the Bivalvia: a systematic and bibliographic catalogue. Bull Am Paleontol 51(232):103–111 + 112–394

Vokes HE (1980) Genera of the Bivalvia: a systematic and bibliographic catalogue (Revised and Updated). Paleontological Research Institute, Ithaca, NY

Von Cosel R (1973) Lista preliminar de los moluscos de la Ciénaga Grande de Santa Marta (Colombia). Boletín de Investigaciones Marinas y Costeras 7:47–56

Von Cosel R (1995) Fifty-one new species of marine bivalves from tropical West Africa. Iberus 13:1–115

von Cosel R, Gofas S (2018) Description of a new genus and twelve new species of marine bivalves from tropical West Africa, with comments on other taxa from the area. Iberus 36(1):1–54

von Ihering H (1907) Les Mollusques fossiles du Tertiaire et du Cretace superieur de l'Argentine. Anales del Museo Nacional de Buenos Aires 14:1–611

Vongpanich V (2000) Family Mactridae (Mollusca Bivalvia) in Thai waters. Phuket Mar Biol Cent Spec Publ 21:483–498

Wakida-Kusunoki A, MacKenzie L (2004) *Rangia* and Marsh Clams, *Rangia cuneata*, *R. flexuosa*, and *Polymesoda caroliniana*, in Eastern México: distribution, biology and ecology, and historical fisheries. Mar Fish Rev 66:13–20

Walckenaer CA (1833) Mémoire sur une nouvelle classification des Aranédes. Société Entomologique de France, Annales 2(3):414–446

Walker RL, Heffernan PB (1994) Age, growth rate, and size of the southern surf clam, *Spisula solidissima similis* (Say, 1822). J Shellfish Res 13:433–441

Walker RL, O'Beirn FX (1996) Embryonic and larval development of *Spisula solidissima similis* (Say, 1822) (Bivalvia: Mactridae). Veliger 39(1):60–64

Warmke G, Abbott RT (1961) Caribbean seashells. A guide to the marine mollusks of Puerto Rico and other West Indian islands, Bermuda and the lower Florida Keys. Livingston Publishing Co, Narberth, PA

Webb CM (1987) Post-larval development of the bivalves *Nucula turgida*, *Venus striatula*, *Spisula subtruncata* and *S. elliptica* (Mollusca: Bivalvia), (with reference to the late larva). J Mar Biol Assoc U K 67(2):441–459

Weinkauff HC (1879, 1880, 1881, 1882, 1884) Die Gattung Mactra. Systematisches Conchylien-Cabinet von Martini und Chemnitz 11(2) [(2)]:1–124. [p. 1–8, pls. 1–6 (1879); p. 9–36, pls. 7–12 (1880); p. 37–68, pls. 13–24 (1881); p. 69–92, pls. 25–30 (1882); p. 93–124, pls. 31–36 (1884)]

Weisbord NE (1964) Late Cenozoic pelecypods from northern Venezuela. Bull Am Paleontol 45(204):5–548

Weisenberg RC (1972) Changes in the organization of tubulin during meiosis in the eggs of the surf clam, *Spisula solidissima*. J Cell Biol 54(2):266–278

Wendell F, DeMartini JD, Dinnel P, Siecke J (1976) The ecology of the gaper or horse clam, *Tresus capax* (Gould 1850) (Bivalvia: Mactridae), in Humboldt Bay, California. Calif Fish Game 62(1):41–64

Wiese L, Niehus O, Faass B, Wiese V (2016) Ein weiteres Vorkommen von *Rangia cuneata* in Deutschland (Bivalvia: Mactridae). Schriften zur Malakozoologie 29:53–60

Wilkins GL (1957) The Cracherode Shell collection. Bull Br Mus Nat Hist Hist Ser 1:121–184

Williams AB, Porter HJ (1971) A ten-year study of meroplankton in North Carolina estuaries: occurrence of postmetamorphal bivalves. Chesap Sci 12(1):26–32

Wilson BR, Kendrick GW (1968) The recent appearance of *Notospisula trigonella* (Lamarck) (Mollusca: Bivalvia: Mactridae) in the Swan estuary. J Malacol Soc Aust 1:25–31

Wong HW (2009) The Mactridae (Mollusca: Bivalvia) of East Coast Park, Singapore. Nat Singapore 2:283–296

Wood W (1818) Index testaceologicus, an illustrated catalogue of British and foreign shells, arranged according to the Linnean system, with the Latin and English names, References to figures and places where found. Printed for W. Wood, London

Wood W (1828) Supplement to the Index testaceologicus, an illustrated catalogue of British and foreign shells. Illustrated with 480 figures. Printed for W. Wood, London

Wood W, Hanley SCT (1856) Index testaceologicus, an illustrated catalogue of British and foreign shells. A new and entirely Revised Edition, with ancient and modern appellations, synonyms, localities, etc. etc. Willis and Sotheran, London

Woodring WP (1982) Geology and paleontology of canal zone and adjoing parts of Panama: description of Tertiary mollusks (Pelecypods: Propeamussiidae to Cuspidariidae). U S Geol Surv Prof Pap 306(F):iv + 541–iv + 759

Woodring WP, Bramlette MN, Kew WSW (1946) Geology and paleontology of Palos Verdes Hills, California. U S Geol Surv Prof Pap 207:1–145

Woodward SP (1856) A manual of the Mollusca; or Rudimentary treatise of recent and fossil shells, vol 3. J. Weale, London

Yancey RM, Welch WR (1968) The Atlantic coast surf clam with a partial bibliography. United States, Department of the Interior, Fish and Wildlife Service, Bureau of Commercial Fisheries, Circular 288

Yonge CM (1948) Cleansing mechanisms and the function of the fourth pallial aperture in *Spisula subtruncata* (da Costa) and *Lutraria lutraria* (L.). J Mar Biol Assoc U K 27(3):585–596

Yonge CM (1957) Mantle fusion in the Lamellibranchia. Pubblicazione della Stazione Zoologica di Napoli 29:151–171

Yonge CM (1982a) Ligamental structure in Mactracea and Myacea (Mollusca: Bivalvia). J Mar Biol Assoc U K 62(1):171–186

Yonge CM (1982b) Mantle margins with a revision of siphonal types in the Bivalvia. J Molluscan Stud 48:102–103

Yonge CM, Allen JA (1985) On significant criteria in establishment of Superfamilies in the Bivalvia: the creation of the superfamily Mesodesmatacea. J Molluscan Stud 51:345–349

Zelaya DG (2016) Marine bivalves from the Argentine coast and continental shelf: species diversity and assessment of the historical knowledge. Am Malacol Bull 33(2):245–262

Zhang Z, Campbell A (2002) Application of a stochastic spawning stock biomass per recruit model for the horse clam fishery in British Columbia. Fish Res 57(1):9–23

Zhidkova LS, Beveh VE, Il'ina AP, Krishtofovich LV, Neverova TI, Savitskii VO, Sheremetjeva GH (1972) Atlas Neogenovykh Molliuskov Kuril'skikh Ostrovov. [Atlas of the Neogene Mollusks of the Kurile Islands]. Akademija Nauk SSSR, Sibirskoe Otdelenie, Sakhalinskii Kompleksnyi, Nachno-Issledovatel'skii Institut. 1–166

Zhuang Q (1983) Two new species of Mactridae (Mollusca Bivalvia) off the Chinese coast. Oceanol Limnol Sin 14(1):88–91

Zilch A (1954) Moluscos de los manglares de El Salvador: con una lista de todas las especies marítimas conocidas de estas regiones: resultados del viaje de exploración del Dr. Adolf Zilch en 1951. Comunicaciones 3(2–3):77–87

Zim HS, Ingle L (1955) Seashores. A guide to animals and plants along the beaches. Golden Nature Guide. Simon et Schuster, New York

Index

J. H. Signorelli, *The Superfamily Mactroidea (Mollusca:Bivalvia) in American Waters*, https://doi.org/10.1007/978-3-030-29097-9

T

The manufacturer's authorised representative in the EU is Springer Nature Customer Service Centre GmbH, Europaplatz 3, 69115 Heidelberg, Germany. If you have any concerns regarding our products, please contact ProductSafety@springernature.com

Printed and bound by CPI Group (UK) Ltd, Croydon, CR0 4YY
30/09/2026
02225665-0007